Introduction to the Classical Theory of Abelian Functions

Recent Titles in This Series

95 **Guangchang Dong,** Nonlinear partial differential equations of second order
94 **Yu. S. Il′yashenko,** Finiteness theorems for limit cycles
93 **A. T. Fomenko and A. A. Tuzhilin,** Elements of the geometry and topology of minimal surfaces in three-dimensional space
92 **E. M. Nikishin and V. N. Sorokin,** Rational approximations and orthogonality
91 **Mamoru Mimura and Hirosi Toda,** Topology of Lie groups, I and II
90 **S. L. Sobolev,** Some applications of functional analysis in mathematical physics, third edition
89 **Valeriĭ V. Kozlov and Dmitriĭ V. Treshchëv,** Billiards: A genetic introduction to the dynamics of systems with impacts
88 **A. G. Khovanskiĭ,** Fewnomials
87 **Aleksandr Robertovich Kemer,** Ideals of identities of associative algebras
86 **V. M. Kadets and M. I. Kadets,** Rearrangements of series in Banach spaces
85 **Mikio Ise and Masaru Takeuchi,** Lie groups I, II
84 **Đáo Trọng Thi and A. T. Fomenko,** Minimal surfaces, stratified multivarifolds, and the Plateau problem
83 **N. I. Portenko,** Generalized diffusion processes
82 **Yasutaka Sibuya,** Linear differential equations in the complex domain: Problems of analytic continuation
81 **I. M. Gelfand and S. G. Gindikin, Editors,** Mathematical problems of tomography
80 **Junjiro Noguchi and Takushiro Ochiai,** Geometric function theory in several complex variables
79 **N. I. Akhiezer,** Elements of the theory of elliptic functions
78 **A. V. Skorokhod,** Asymptotic methods of the theory of stochastic differential equations
77 **V. M. Filippov,** Variational principles for nonpotential operators
76 **Phillip A. Griffiths,** Introduction to algebraic curves
75 **B. S. Kashin and A. A. Saakyan,** Orthogonal series
74 **V. I. Yudovich,** The linearization method in hydrodynamical stability theory
73 **Yu. G. Reshetnyak,** Space mappings with bounded distortion
72 **A. V. Pogorelev,** Bendings of surfaces and stability of shells
71 **A. S. Markus,** Introduction to the spectral theory of polynomial operator pencils
70 **N. I. Akhiezer,** Lectures on integral transforms
69 **V. N. Saliĭ,** Lattices with unique complements
68 **A. G. Postnikov,** Introduction to analytic number theory
67 **A. G. Dragalin,** Mathematical intuitionism: Introduction to proof theory
66 **Ye Yan-Qian,** Theory of limit cycles
65 **V. M. Zolotarev,** One-dimensional stable distributions
64 **M. M. Lavrent′ev, V. G. Romanov, and S. P. Shishat·skiĭ,** Ill-posed problems of mathematical physics and analysis
63 **Yu. M. Berezanskiĭ,** Selfadjoint operators in spaces of functions of infinitely many variables
62 **S. L. Krushkal′, B. N. Apanasov, and N. A. Gusevskiĭ,** Kleinian groups and uniformization in examples and problems
61 **B. V. Shabat,** Distribution of values of holomorphic mappings
60 **B. A. Kushner,** Lectures on constructive mathematical analysis
59 **G. P. Egorychev,** Integral representation and the computation of combinatorial sums
58 **L. A. Aĭzenberg and A. P. Yuzhakov,** Integral representations and residues in multidimensional complex analysis
57 **V. N. Monakhov,** Boundary-value problems with free boundaries for elliptic systems of equations

(See AMS catalogue for earlier titles.)

Translations of

MATHEMATICAL MONOGRAPHS

Volume 96

Introduction to the Classical Theory of Abelian Functions

A. I. Markushevich

American Mathematical Society
Providence, Rhode Island

ВВЕДЕНИЕ В КЛАССИЧЕСКУЮ ТЕОРИЮ АБЕЛЕВЫХ ФУНКЦИЙ

МОСКВА «НАУКА» 1979

Translated from the Russian by G. Bluher
Translation edited by Ralph P. Boas

1991 *Mathematics Subject Classification*. Primary 14K20; Secondary 33E05.

ABSTRACT. Introduction to the classical theory of Abelian functions. Moscow, Nauka, Glavnaya redaktsiya fisiko-matematicheskoy literaturi, 1979, 240 pages.

The theory of Abelian functions, which was at the center of mathematics of the XIX century, is attracting a significant amount of attention again. But today it is frequently considered not as a chapter of the general theory of functions but as an area of application of the ideas and methods of commutative algebra. In this book, we present an exposition of the fundamentals of the theory of Abelian functions based on the methods of the classical theory of functions. This theory includes the theory of elliptic functions as a special case. A detailed historical introduction also distinguishes our exposition from others.

This book is directed toward undergraduate and graduate students and toward the mathematicians teaching at universities. The reader is assumed to have acquired only the minimal set of facts about functions of many variables. Such facts are usually presented in a standard course on the theory of analytic functions given at universities. All additional necessary material is given in the Appendices. The book is written on the basis of a course which was taught by the author in the Fall semester of the 1976/77 academic year in the Department of Professional Development (Fakultet Povisheniya Kvalifikatsii) of Moscow State Univeristy.

Library of Congress Cataloging-in-Publication Data

Markushevich, A. I. (Alekseĭ Ivanovich), 1908–
[Vvedenie v klassicheskuiu teoriiu Abelevykh funktsii. English]
Introduction to the classical theory of Abelian functions/A. I. Markushevich; [translated from the Russian by G. Bluher; translation edited by Ralph P. Boas]
p. cm.—(Translations of mathematical monographs; v. 96)
Translation of: Vvedenie v klassicheskuiu teoriiu Abelevykh funktsii.
ISBN 0-8218-4542-X (alk. paper)
1. Functions, Abelian. I. Boas, Ralph Philip. II. Title. III. Series.
QA345.M4513 1992 91-36838
515—dc20 CIP

Printed in the United States of America

Information on Copying and Reprinting can be found at the back of this volume.

The paper used in this book is acid-free and falls within the guidelines
established to ensure permanence and durability.
This publication was typeset using $\mathcal{A}\mathcal{M}\mathcal{S}$-TEX,
the American Mathematical Society's TEX macro system.

10 9 8 7 6 5 4 3 2 1 97 96 95 94 93 92

Contents

Preface

"When I was a student, during the First World War," recalls F. Klein, "Abelian functions were considered (under the influence inherited from Jacobi) to be the indisputable pinnacle of mathematics. Each of us, naturally, had some ambition to make an advance in this area on his own. But now? It is doubtful whether the younger generation is familiar with Abelian functions at all."([1])

More than half a century has passed since the time of this quotation and the interest in the "temporarily outdated" (as Klein called it) theory of Abelian functions is again on the rise. However, today it has become quite different from the theory that Klein was discussing. It is considered not as a chapter of the theory of functions, but rather as a chapter or even an area of application of the ideas and methods of commutative algebra, algebraic geometry, and complex analysis. A typical example of such interpretations is the book "Introduction to Algebraic and Abelian Functions" by the algebraist S. Lang.([2]) We may also mention several books, for example: "Introduction à l'étude des variétés kahleriennes" by André Weil([3]) (mainly Chapter VI) and "Abelian Varieties" by D. Mumford([4]) (in collaboration with K. P. Ramanujan).

With all due respect to these excellent monographs, we still would like to offer the reader a different exposition of the subject. Our exposition is based on the ideas and methods of the classical theory of functions. Here the reader is assumed to know only a minimal number of facts about functions of several variables, such as are commonly presented in a standard university course on the theory of analytic functions. All the additional material that will be needed is given in the appropriate parts of this book.

Our exposition is distinguished by a detailed historical introduction (Chapter I), covering the period up to the middle of the nineteenth century during which the initial accumulation of the ideas and facts took place. In the next

---

([1])F. Klein, *Vorlesungen über die Entwicklung der Mathematik im* 19. *Jahrhundert*, Part I, Springer, 1926.

([2])S. Lang, *Introduction to algebraic and Abelian functions*, Addison-Wesley, 1972.

([3])A. Weil, *Introduction à l'étude des variétés kähleriennes*, Actualités Sci. Indust., no. 1267, Hermann, Paris, 1958.

([4])D. Mumford, *Abelian Varieties*, Oxford Univ. Press, 1970.

three chapters we present a systematic construction of the foundations of the theory of Abelian functions as considered as a study of meromorphic functions of p complex variables with the maximum number ($2p$) of independent periods. This theory includes as a special, and in some sense trivial, case the theory of elliptic functions ($p = 1$). In the three main chapters we have been considerably influenced by Fabio Conforto's book "Abelsche Functionen und algebraische Geometry," edited by W. Gröbner, A. Andreotti, and M. Rosati.([5]) This book was published in 1956 in the series "Grundlagen der mathematischen Wissenschaften" in which Klein's lectures also appeared. We note that an exposition which is similar in spirit can be found in C. L. Siegel's book "Analytic functions of several variables"([6]) (Chapters IV-IX). The contents of this book which we now offer to readers (undergraduate and graduate students, as well as to university teachers of mathematics) was the subject of a course given in the fall semester of 1976/77 in the Department of Professional Development of University Instructors at the Moscow State University.

In conclusion I thank my daughter L. A. Markushevich, who assisted me in the preparation of the manuscript for publication.

A. Markushevich

([5])See footnote ([18]) on page 44.

([6])C. L. Siegel, *Analytic functions of several complex variables*, Institute for Advanced Study, Princeton, N. J., 1950; Russian transl., IL, Moscow, 1954.

CHAPTER I

Historical Introduction. The Jacobi Inversion Problem

§1. Euler's equation

Like a mighty river that starts as a modest, hardly noticeable, spring, a scientific theory is generated by elementary, apparently commonplace facts. In any case, if we start from the contemporary state of the theory of Abelian functions, which algebraists now feel that they completely own, and if we continue against the flow of time and trace the origins of the theory, we find ourselves in the middle of the 18th century, considering two familiar aspects of integral calculus:

(A) the inverse function of the integral

$$x = \int_0^u \frac{dt}{\sqrt{1-t^2}},$$

is single-valued and entire ($u = \sin x$) and has period

$$2\pi = 2\int_{-1}^1 \frac{dt}{\sqrt{1-t^2}};$$

(B) the sum of the two integrals

$$\int_0^u \frac{dt}{\sqrt{1-t^2}} + \int_0^v \frac{dt}{\sqrt{1-t^2}}$$

can be expressed as a single integral

$$\int_0^w \frac{dt}{\sqrt{1-t^2}},$$

where the upper limit is an algebraic function of u and v

$$w = u\sqrt{1-v^2} + v\sqrt{1-u^2}.$$

The last statement is evidently nothing but the addition theorem for the sine function: if $\sin x = u$ and $\sin y = v$, then

$$w = \sin(x+y) = u\sqrt{1-v^2} + v\sqrt{1-u^2}.$$

We may find a source for the development of the theory of Abelian functions in the following question: *to what degree are the preceding relations determined by the fact that the integral has the quadratic polynomial* $1 - t^2$ *under the radical sign? Is it possible, instead, to use a polynomial of higher degree?* Euler does not state this question explicitly. However, in book 1, part 1, section 2 of his "Integral calculus" (see the modern Russian edition, v.1, pp. 324–376) we find a significant result (first obtained by him in the '50s of the 18th century) which demonstrates that at least property (B) is preserved if one uses a quartic polynomial instead of a quadratic polynomial.

Here is his argument, with some simplifications that are appropriate for our purposes.

First consider an algebraic equation of the form

$$u^2 + v^2 + Au^2v^2 + 2Buv - C^2 = 0, \tag{I.1.1}$$

where A, B, and C are constants. After differentiating, we obtain

$$[u(1 + Av^2) + Bv]\,du + [v(1 + Au^2) + Bu]\,dv = 0. \tag{I.1.2}$$

But

$$(1 + Av^2)u^2 + 2Buv + v^2 - C^2 = 0,$$

and therefore

$$u = -\frac{Bv}{1 + Av^2} + \frac{\sqrt{B^2v^2 + (C^2 - v^2)(1 + Av^2)}}{1 + Av^2},$$

i.e.,

$$u(1 + Av^2) + Bv = \sqrt{C^2 + (AC^2 + B^2 - 1)v^2 - Av^4}. \tag{I.1.3}$$

Since equation (I.1.1) is symmetric in u and v, we can interchange u and v in (I.1.3) to obtain

$$v(1 + Au^2) + Bu = \sqrt{C^2 + (AC^2 + B^2 - 1)u^2 - Au^4}. \tag{I.1.4}$$

By (I.1.3) and (I.1.4), equation (I.1.2) becomes

$$\frac{du}{\sqrt{C^2 + (AC^2 + B^2 - 1)u^2 - Au^4}} + \frac{dv}{\sqrt{C^2 + (AC^2 + B^2 - 1)v^2 - Av^4}} = 0. \tag{I.1.5}$$

It then follows that (I.1.1) is a solution of the differential equation (I.1.5). In order to reduce this equation to the form

$$\frac{du}{\sqrt{1 - u^2}} + \frac{dv}{\sqrt{1 - v^2}} = 0, \tag{I.1.6}$$

we need only set $A = 0$, $B^2 = 1 - C^2$. Then the equation (I.1.1) becomes

$$u^2 + v^2 + 2\sqrt{1 - C^2}\,uv - C^2 = 0.$$

We transform this to the form $\Phi(u, v) = C$, i.e., solve it for C. We get successively

$$C^4 - 2(u^2 + v^2 - 2u^2v^2)C^2 + (u^2 - v^2)^2 = 0,$$

$$\begin{aligned} C^2 &= u^2(1 - v^2) + v^2(1 - u^2) + 2uv\sqrt{(1 - u^2)(1 - v^2)}, \\ C &= u\sqrt{1 - v^2} + v\sqrt{1 - u^2}. \end{aligned} \tag{I.1.7}$$

This is an integral of (I.1.6), represented in algebraic form. However, an integral of (I.1.6) can also be immediately written in the transcendental form

$$\int_0^u \frac{dt}{\sqrt{1 - t^2}} + \int_0^v \frac{dt}{\sqrt{1 - t^2}} = \alpha + \beta = \gamma = \text{const}. \tag{I.1.8}$$

Relation (I.1.7) must follow from (I.1.8). If we set $u = 0$ in (I.1.8), then we obtain

$$\int_0^v \frac{dt}{\sqrt{1 - t^2}} = \gamma,$$

i.e., $v = \sin \gamma$; on the other hand, (I.1.7) implies that $v = C$. Hence

$$C = \sin \gamma = \sin(\alpha + \beta), \qquad \gamma = \int_0^C \frac{dt}{\sqrt{1 - t^2}}$$

and finally

$$\int_0^u \frac{dt}{\sqrt{1 - t^2}} + \int_0^v \frac{dt}{\sqrt{1 - t^2}} = \int_0^{u\sqrt{1-v^2}+v\sqrt{1-u^2}} \frac{dt}{\sqrt{1 - t^2}}. \tag{I.1.9}$$

Thus, Euler rederived the addition theorem for the sine by using integral calculus. His method allowed him to obtain an analogous relation for a quartic polynomial. However, he did not raise the question of a function inverse to the corresponding integral. That step was only to be taken three decades later by the young Gauss in his mathematical diary.

For the time being, following Euler, we rewrite (I.1.5) as

$$\frac{du}{\sqrt{1 + mu^2 + nu^4}} + \frac{dv}{\sqrt{1 + mv^2 + nv^4}} = 0. \tag{I.1.10}$$

To accomplish this we have only to set $AC^2 + B^2 - 1 = mC^2$, $A = -nC^2$, i.e., $B^2 = 1 + mC^2 + nC^4$. Equation (I.1.1) now becomes

$$u^2 + v^2 - nC^2u^2v^2 + 2\sqrt{1 + mC^2 + nC^4}uv - C^2 = 0.$$

With the abbreviation $1 + mt^2 + nt^4 = P(t)$, we find successively

$$\begin{gathered} 4(1 + mC^2 + nC^4)u^2v^2 = [(u^2 + v^2) - (nu^2v^2 + 1)C^2]^2, \\ (1 - nu^2v^2)^2C^4 - 2[u^2P(v) + v^2P(u)]C^2 + (u^2 - v^2) = 0; \end{gathered}$$

or, if we notice that $(1-nu^2v^2)(u^2-v^2) = u^2P(v) - v^2P(u)$, we obtain

$$C^2 = \frac{u^2P(v) + v^2P(u) + 2uv\sqrt{P(u)P(v)}}{(1-nu^2v^2)^2},$$

from which we finally obtain

$$C = \Phi(u\,,\,v) = \frac{u\sqrt{P(v)} + v\sqrt{P(u)}}{1-nu^2v^2}. \tag{I.1.11}$$

This is an integral of the Euler equation (I.1.10) expressed in algebraic form.

If we compare this in the same way with the transcendental form of the equation

$$\int_0^u \frac{dt}{\sqrt{P(t)}} + \int_0^v \frac{dt}{\sqrt{P(t)}} = \alpha + \beta = \gamma = \text{const},$$

we obtain the relation

$$\int_0^u \frac{dt}{\sqrt{P(t)}} + \int_0^v \frac{dt}{\sqrt{P(t)}} = \int_0^{\Phi(u\,,\,v)} \frac{dt}{\sqrt{P(t)}}. \tag{I.1.12}$$

This extends property (B) to the case of a quartic polynomial $P(t) = 1 + mt^2 + nt^4$.

§2. Elliptic functions of Gauss

Young Gauss (in 1796, at the latest) became interested in the elliptic integral $\int_0^u dt/\sqrt{1-t^4}$ because it is the closest analog of $\int_0^u dt/\sqrt{1-t^2}$. The first integral plays the same role with respect to Bernoulli's lemniscate $r^2 = \sin(2\theta)$ as the second with respect to the circle $r = \sin\theta$. Guided by this analogy, Gauss accomplished the inversion of the elliptic integral and thus was the first to discover elliptic functions and their most important properties. Unfortunately, he did not publish his results in this field, and so the credit for being the founders of the theory of elliptic functions has been ascribed to Abel and Jacobi.

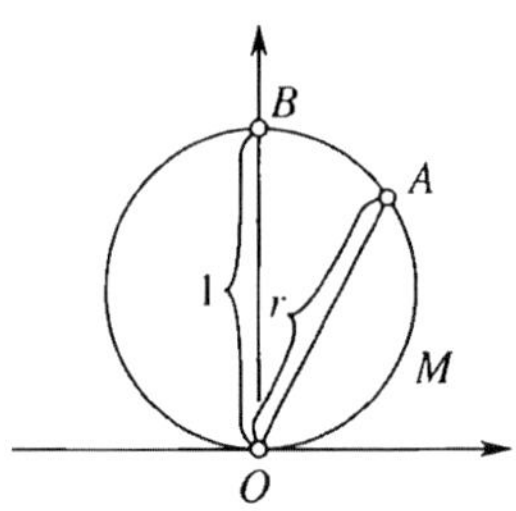

FIGURE 1

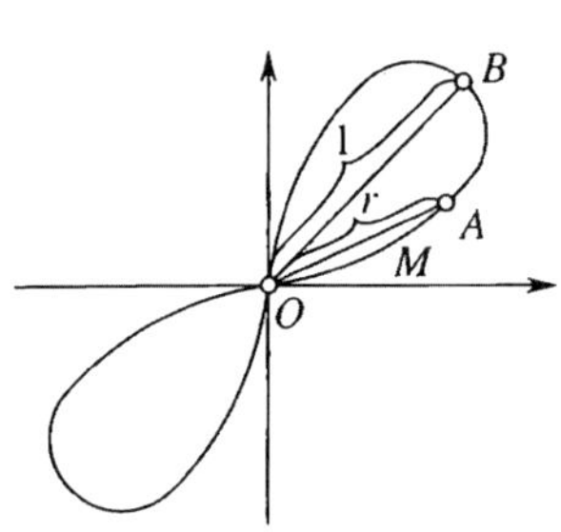

FIGURE 2

The very simple and natural flow of Gauss's reasoning can best be reconstructed by means of the following comparisons using Figures 1 and 2

Trigonometric functions	Lemniscate functions
$\overset{\frown}{OMA} = t = \int_0^r \frac{dr}{\sqrt{1-t^2}}$	$\overset{\frown}{OMA} = t = \int_0^r \frac{dr}{\sqrt{1-r^4}}$
$r = \sin t$	$r = \operatorname{sl} t$ (lemniscate sine)
$\overset{\frown}{OMB} = \int_0^1 \frac{dr}{\sqrt{1-r^2}}$	$\overset{\frown}{OMB} = \int_0^1 \frac{dr}{\sqrt{1-r^4}}$
$= \frac{\pi}{2} (\approx 1.5708)$	$= \frac{\varpi}{2} (\approx 1.3110)$
$\cos t = \sin\left(\frac{\pi}{2} - t\right)$	$\operatorname{cl} t = \operatorname{sl}(\varpi/2 - t)$ (lemniscate cosine)
$\sin\frac{\pi}{2} = 1$, $\sin \pi = 0$	$\operatorname{sl}\frac{\varpi}{2} = 1$, $\operatorname{sl}\varpi = 0$
$\sin(\pi + t) = -\sin t$,	$\operatorname{sl}(\varpi + t) = -\operatorname{sl} t$,
$\sin(-t) = -\sin t$	$\operatorname{sl}(-t) = -\operatorname{sl} t$.

Euler's addition theorems

$\sin(\tau + t) = \sin\tau\sqrt{1-\sin^2 t} + \sin t\sqrt{1-\sin^2\tau}$	$\operatorname{sl}(\tau + t) = \dfrac{\operatorname{sl}\tau\sqrt{1-\operatorname{sl}^4 t} + \operatorname{sl} t\sqrt{1-\operatorname{sl}^4\tau}}{1 + \operatorname{sl}^2\tau\operatorname{sl}^2 t}$ (corollary of (I.1.12) when $P(t) = 1 - t^4$)

Replacing τ by $\pi/2$ and by $\varpi/2$, and $-t$ by t, we get

$\cos t = \sin\left(\frac{\pi}{2} - t\right) = \sqrt{1 - \sin^2 t}$	$\operatorname{cl} t = \operatorname{sl}\left(\frac{\varpi}{2} - t\right) = \sqrt{\frac{1-\operatorname{sl}^2 t}{1+\operatorname{sl}^2 t}}$
$1 = \sin^2 t + \cos^2 t$	$1 = \operatorname{sl}^2 t + \operatorname{cl}^2 t + \operatorname{sl}^2 t \operatorname{cl}^2 t$

Functions of purely imaginary argument

Setting $x = iy$, where y is real, we obtain

$\int_0^{ri} \frac{dx}{\sqrt{1-x^2}} = i\int_0^r \frac{dy}{\sqrt{1+y^2}} = it$	$\int_0^{ri} \frac{dx}{\sqrt{1-x^4}} = i\int_0^r \frac{dy}{\sqrt{1-y^4}} = it$
$ri = \sin(it)$, $t = \ln(\sqrt{1+r^2} + r)$	$ri = \operatorname{sl}(it)$, $r = \operatorname{sl} t$
$r = \sinh t$	
$\sin(it) = i\sinh t$, $\cos(it) = \cosh t$	$\operatorname{sl}(it) = i\operatorname{sl} t$, $\operatorname{cl}(it) = \frac{1}{\operatorname{cl} t}$.

(the last formula uses an algebraic relation between the lemniscate sine and cosine).

Definition of functions of a complex variable
by means of an addition theorem

$\sin(x+iy) = \sin x \cosh y + i\cos x \sinh y$	$\mathrm{sl}(x+iy) = \frac{\mathrm{sl}\,x\sqrt{1-\mathrm{sl}^4 y}+i\,\mathrm{sl}\,y\sqrt{1-\mathrm{sl}^4 x}}{1-\mathrm{sl}^2 x\,\mathrm{sl}^2 y} = \frac{\mathrm{sl}^2 x+\mathrm{sl}^2 y}{\mathrm{sl}\,x\sqrt{1-\mathrm{sl}^4 y}-i\,\mathrm{sl}\,y\sqrt{1-\mathrm{sl}^4 x}}.$

Periods

$\Omega = 2n\pi \ (n=0, \pm 1, \pm 2, \dots)$	$\Omega = 2m\varpi + 2n\varpi i$, $(m, n = 0, \pm 1, \pm 2, \dots)$

Zeros

$n\pi$	$(m+ni)\varpi$

Poles

None	$[(2m+1)+(2n+1)i]\frac{\varpi}{2}.$

A detailed development of the analogy between trigonometric and lemniscate functions was given by the author in his popular expository booklet "Wonderful sines." (Nauka, Moscow, 1975).

Gauss, however, was not content to stop at this point and went on to develop an analytic apparatus for representing the new class of functions.

The expansion of the sine as an infinite product was considered by Euler in his first article (in "Kommentarii Imperatorskoĭ Peterburgskoĭ Akademii Nauk", 1734–1735, published in 1740), where it was presented as a natural generalization of the factorization of a polynomial. If a polynomial $P(x)$ with zeros

$$0, \pm a, \pm b, \pm c, \dots$$

is represented in the form

$$P(x) = Cx\left(1-\frac{x^2}{a^2}\right)\left(1-\frac{x^2}{b^2}\right)\left(1-\frac{x^2}{c^2}\right)\cdots,$$

where $C = \lim_{x\to 0} P(x)/x$, then $\sin x$ (whose zeros we know) must be representable in the form

$$\sin x = Cx\left(1-\frac{x^2}{\pi^2}\right)\left(1-\frac{x^2}{4\pi^2}\right)\left(1-\frac{x^2}{9\pi^2}\right)\cdots,$$

where

$$C = \lim_{x\to 0}\frac{\sin x}{x} = 1.$$

Apparently, at least in the beginning, Gauss was satisfied with this reasoning. Thus, after finding all the zeros and poles of $\mathrm{sl}\,z$ $(z = x+iy)$, he expresses $\mathrm{sl}\,z$ in the form of a quotient of two infinite products in which zeros of the numerator are the zeros of $\mathrm{sl}\,z$ and the zeros of the denominator are the poles of $\mathrm{sl}\,z$:

$$\mathrm{sl}\,z = \frac{M(z)}{N(z)} = \frac{z\prod'\left[1-\frac{z}{(m+ni)\varpi}\right]}{\prod\left\{1-\frac{2z}{[(2m+1)+(2n+1)i]\varpi}\right\}};$$

where the prime $'$ means, as usual, that m and n in the product in the numerator are not simultaneously zero. From our point of view both products are divergent. However, they can nevertheless be used if we understand them in the sense of principal values (as, for example, limits of finite products defined by the conditions $|m| \leq N$, $|n| \leq N$ as $N \to \infty$). Similar products and series were used by Eisenstein half a century after Gauss developed these formulas. Today the use of these products and series is being revived by A. Weil ([7]) who considers them to be quite productive. It follows from this representation that $\operatorname{sl} z$ is a quotient of two entire functions and therefore is a *meromorphic function.*

Gauss also found another representation for $\operatorname{sl} z$ that has the advantage of converging absolutely and even rapidly. This step is very important for our purposes since it led to the appearance of *theta functions* in Gauss's work, and since a generalization of theta functions is a basis for the theory of Abelian functions.

In order to derive the new representation of $\operatorname{sl} z$, Gauss uses the change of variable $\zeta = \sin(\pi z/\varpi)$. The zeros of $\operatorname{sl} z$ for the transformed function $\operatorname{sl}[z(\zeta)]$ at the points become

$$\sin[(m \pm ni)\pi] = \pm i \sinh n\pi = \pm i\frac{e^{n\pi} - e^{-n\pi}}{2}, \qquad n = 0, 1, \dots,$$

and the poles become

$$\sin\left\{[(2m \pm 1) + (2n+1)i]\frac{\pi}{2}\right\}$$
$$= \pm\cosh\frac{(2n+1)\pi}{2} = \pm\frac{e^{\frac{2n+1}{2}\pi} + e^{-\frac{2n+1}{2}\pi}}{2}, \qquad n = 0, 1, 2, \dots.$$

Thus, two-dimensional point lattices of zeros and poles are transformed into two sequences of points on the imaginary and real axes. This leads Gauss to propose the following formula for representing the function $\operatorname{sl} z$:

$$\operatorname{sl} z = \frac{P(z)}{Q(z)} = \frac{(\varpi/2)\zeta\prod_1^\infty[1 + 4\zeta^2/(e^{n\pi} - e^{-n\pi})^2]}{\prod_1^\infty[1 - 4\zeta^2/(e^{(2n+1)\pi/2} + e^{-(2n+1)\pi/2})^2]},$$
$$\zeta = \sin\frac{\pi}{\varpi}z,$$

where it is obvious that the products converge very rapidly.

Finally, he was able to go further and express the functions $P(z)$ and $Q(z)$ as Fourier series (by expressing the products as power series and substituting sines and cosines of integral multiples of $z\pi/\varpi$ for powers of ζ). Gauss sets $z = \psi\varpi$ and writes these series in their final form as

$$P(\psi\varpi) = 2^{3/4}\sqrt{\frac{\pi}{\varpi}}(e^{-\pi/4}\sin\psi\pi - e^{-9\pi/4}\sin 3\psi\pi + \cdots),$$

$$Q(\psi\varpi) = 2^{-1/4}\sqrt{\frac{\pi}{\varpi}}(1 + 2e^{-\pi}\cos 2\psi\pi + 2e^{-4\pi}\cos 4\psi\pi + \cdots).$$

([7]) A. Weil, *Elliptic functions according to Eisenstein and Kronecker*, Springer-Verlag, 1976.

Thus

$$\operatorname{sl} z = 2e^{-\pi/4}\frac{\sin\psi\pi - e^{-2\pi}\sin 3\psi\pi + \cdots}{1 + 2e^{-\pi}\cos 2\psi\pi + 2e^{-4\pi}\cos 4\psi\pi + \cdots}. \tag{I.2.1}$$

This, then, is a representation of the elliptic function $\operatorname{sl} z$ as a quotient of two *theta functions*. The general definition of theta functions will be given later.

In 1825–1830, Abel and Jacobi followed essentially the same path as Gauss, independently of Gauss and of each other. However, they immediately inverted an elliptic integral more general than the lemniscate integral. (This, by the way, had also been done by Gauss in 1798–99, i.e., long before Abel and Jacobi.) In fact, Abel starts with the integral

$$\alpha = \int_0^x \frac{dx}{\sqrt{(1-c^2x^2)(1+e^2x^2)}},$$

sets $x = \varphi(\alpha)$ and also introduces the functions

$$f(\alpha) = \sqrt{1 - c^2\varphi^2(\alpha)}, \qquad F(\alpha) = \sqrt{1 + e^2\varphi^2(\alpha)}.$$

Jacobi, on the other hand, first inverts the integral $\int_0^\varphi d\varphi/\sqrt{1-k^2\sin^2\varphi} = u$ and calls φ the *amplitude* of u.

Then Jacobi makes the substitution $x = \sin\varphi$ for an elliptic integral of the first kind (in the Legendre normal form)

$$\int_0^x \frac{dx}{\sqrt{(1-x^2)(1-k^2x^2)}} = u;$$

as a result, he obtains

$$\int_0^\varphi \frac{d\varphi}{\sqrt{1-k^2\sin^2\varphi}} = u, \qquad \varphi = \operatorname{am} u,$$

and $x = \sin(\operatorname{am} u)$ (sine of the amplitude of u) is the principal Jacobian *elliptic function*. In addition, Jacobi introduces two other elliptic functions

$$\cos\operatorname{am} u = \sqrt{1 - \sin\operatorname{am}^2 u}, \qquad \Delta\operatorname{am} u = \sqrt{1 - k^2\sin\operatorname{am}^2 u}.$$

Since the time of Gudermann (1838) the abbreviated notation $\operatorname{sn} u$, $\operatorname{cn} u$, $\operatorname{dn} u$ has been used for the Jacobian elliptic functions .

§3. Jacobi's inversion method

At the end of his "New foundations of the theory of elliptic functions" (in Latin, Königsberg,1829), Jacobi obtained expressions of the following form for his elliptic functions:

$$\operatorname{sn} u = \frac{1}{\sqrt{k}}\frac{\mathrm{H}(u)}{\Theta(u)}, \qquad \operatorname{cn} u = \sqrt{\frac{k'}{k}}\frac{\mathrm{H}(u+K)}{\Theta(u)},$$

$$\operatorname{dn} u = \sqrt{k'}\,\frac{\Theta(u+K)}{\Theta(u)}.$$

The eta and theta functions $\mathrm{H}(u)$ and $\Theta(u)$ are entire functions with no common zeros and have the following properties of generalized periodicity:

$$\Theta(u+2K)=\Theta(u)\,,\qquad \Theta(u+2iK')=-e^{-\pi iK'/K}e^{-\pi iu/K}\Theta(u)\,,$$

$$\mathrm{H}(u+2K)=-\mathrm{H}(u)\,,\qquad \mathrm{H}(u+2iK')=-e^{-\pi iK'/K}e^{-\pi iu/K}\mathrm{H}(u)\,,$$

where K and K' are nonzero constants (real and positive in Jacobi's papers). Thus,

$$K=\int_0^{\pi/2}\frac{d\varphi}{\sqrt{1-k^2\sin^2\varphi}}\quad\text{and}\quad K'=\int_0^{\pi/2}\frac{d\varphi}{\sqrt{1-k'^2\sin^2\varphi}}$$
$$((k')^2+k^2=1).$$

It follows that the elliptic function $\operatorname{sn}u$ is (in modern terminology) a *doubly periodic meromorphic* function and one can take $4K$ and $2iK'$ to be its fundamental periods.

The idea of the *Jacobi inversion method* was to first construct functions with the necessary properties of generalized periodicity (theta functions) and only then ascertain that appropriately constructed relations among them solve the problem of the inversion of elliptic integrals. A ten-hour expository lecture by Jacobi at Königsberg University (1835–1836) was devoted to the development of this idea. After Jacobi's death his pupil Borchardt published a development of it.

A series of the form:

$$\sum_{n=-\infty}^{\infty}e^{an^2+bn+c}$$

plays a fundamental role, where b and c are linear functions of x, and a is a complex constant with *negative real part* (this condition guarantees the convergence of the series). This series is preferably written in the form introduced by Hermite.

Let

$$\Theta_{\mu,\nu}(x)=\sum\exp\left\{\pi i\left[n\nu+(2n+\mu)x+\tau\frac{(2n+\mu)^2}{4}\right]\right\},\tag{I.3.1}$$

where μ and ν take only real(integral) values and $\tau=\tau'+i\tau''$ is a complex parameter related to the complex number a introduced above by the equation $a=\pi i\tau$. In accordance with what we said above, we restrict τ by the condition

$$\operatorname{Im}\tau=\tau''>0.\tag{I.3.2}$$

Let $x=x'+ix''$. We denote the exponent of the general term of the series (I.3.1) by A_n and prove that if (I.3.2) is satisfied this series converges absolutely and uniformly in each region of the form $|x''|<R$, where R is

an arbitrary positive number. It will follow from this that the sum of the series is an entire function of x.

Indeed,

$$\begin{aligned}\operatorname{Re} A_n &= -\pi(2n+\mu)x'' - \pi\tau''\frac{(2n+\mu)^2}{4} \\ &= -\pi\tau'' n^2\left[1+O\left(\frac{1}{n}\right)\right] < -\frac{1}{2}\pi\tau'' n^2\end{aligned}$$

for all sufficiently large n. Hence

$$|\exp A_n| = \exp(\operatorname{Re} A_n) < [\exp(-\pi\tau''/2)]^{n^2}.$$

Noticing that, under the restriction (I.3.2) satisfied by τ we have $\exp(-\pi\tau''/2) = \alpha < 1$, we have

$$|\exp A_n| < \alpha^{n^2} \le \alpha^{|n|}.$$

From this, the assertion about convergence readily follows. Thus $\Theta_{\mu,\nu}(x)$ is an entire function.

Let $q = e^{\pi i\tau}$, $0 < |q| < 1$, and define in terms of $\Theta_{\mu,\nu}(x)$ the four Jacobi theta functions:

$$\begin{aligned}\vartheta_1(x) &= -i\Theta_{1,1}(x) \\ &= -i\sum_{-\infty}^{+\infty}\exp\left\{\pi i\left[n+(2n+1)x+\tau\frac{(2n+1)^2}{4}\right]\right\} \\ &= -i\sum_{-\infty}^{\infty}(-1)^n q^{(2n+1)^2/4}e^{(2n+1)\pi i x} \\ &= \sum_{0}^{+\infty}(-1)^n 2q^{(2n+1)^2/4}\sin(2n+1)\pi x,\end{aligned} \tag{I.3.3}$$

$$\begin{aligned}\vartheta_2(x) &= \Theta_{1,0}(x) = \sum_{-\infty}^{+\infty}\exp\left\{\pi i\left[(2n+1)x+\tau\frac{(2n+1)^2}{4}\right]\right\} \\ &= \sum_{-\infty}^{+\infty}q^{(2n+1)^2/4}e^{(2n+1)\pi i x} = \sum_{0}^{\infty}2q^{(2n+1)^2/4}\cos(2n+1)\pi x,\end{aligned} \tag{I.3.4}$$

$$\begin{aligned}\vartheta_3(x) &= \Theta_{0,0}(x) = \sum_{-\infty}^{\infty}\exp[\pi i(2nx+\tau n^2)] \\ &= \sum_{-\infty}^{\infty}q^{n^2}e^{2n\pi i x} = 1+\sum_{1}^{\infty}2q^{n^2}\cos 2n\pi x,\end{aligned} \tag{I.3.5}$$

$$\begin{aligned}\vartheta_4(x) &= \Theta_{0,1}(x) = \sum_{-\infty}^{\infty}\exp[\pi i(n+2xn+\tau n^2)] \\ &= \sum_{-\infty}^{\infty}(-1)^n q^{n^2}\exp(2n\pi i x) = 1+\sum_{1}^{\infty}(-1)^n 2q^{n^2}\cos 2n\pi x.\end{aligned} \tag{I.3.6}$$

TABLE 1

	$x+\frac{1}{2}$	$x+\frac{\tau}{2}$	$x+\frac{1}{2}+\frac{\tau}{2}$	$x+1$	$x+\tau$
ϑ_1	ϑ_2	$iM\vartheta_4$	$M\vartheta_3$	$-\vartheta_1$	$-N\vartheta_1$
ϑ_2	$-\vartheta_1$	$M\vartheta_3$	$-iM\vartheta_4$	$-\vartheta_2$	$N\vartheta_2$
ϑ_3	ϑ_4	$M\vartheta_2$	$iM\vartheta_1$	ϑ_3	$N\vartheta_3$
ϑ_4	ϑ_3	$iM\vartheta_1$	$M\vartheta_2$	ϑ_4	$-N\vartheta_4$

$$M = M(x) = q^{-1/4}\exp(-\pi i x)$$
$$N = N(x) = q^{-1}\exp(-2\pi x)$$

It is obvious that, in terms of these definitions, Gauss's formula (I.2.1) for the lemniscate sine can be written in the form

$$\operatorname{sl} z = \frac{\vartheta_1(\psi\pi)}{\vartheta_3(\psi\pi)} \qquad (z = \psi\pi),$$

where the functions ϑ_1 and ϑ_3 are taken with the particular value $q = e^{-\pi}$ (i.e., $\tau = i$).

It follows from the formulas for $\vartheta_j(x)$ that $\vartheta_1(x)$ is an odd function and the rest of the theta functions are *even*.

From the definition of theta functions it is easy to derive formulas expressing the values $\vartheta_j(x+\frac{1}{2})$, $\vartheta_j(x+\frac{\tau}{2})$, $\vartheta_j(x+\frac{1}{2}+\frac{\tau}{2})$, $\vartheta_j(x+1)$, $\vartheta_j(x+\tau)$ in terms of $\vartheta_j(x)$ $(j = 1, 2, 3, 4)$. These results, which are entirely analogous to the *reduction* formulas of elementary trigonometry, are given in Table 1. In the columns of this table, starting where the first, ϑ_j stands for $\vartheta_j(x)$ $(j = 1, 2, 3, 4)$. With the help of Table 1 we conclude that the quotients of the first three by the fourth are *doubly periodic meromorphic functions*. Table 2 shows the periods of the functions $\vartheta_i(x)/\vartheta_4(x)$ $(i = 1, 2, 3)$.

Jacobi, however, was interested in these functions not for their own sake, but for their applicability to the problem of inversion of elliptic integrals.

TABLE 2

	Periods	
$\frac{\vartheta_1}{\vartheta_4}$	2;	τ
$\frac{\vartheta_2}{\vartheta_4}$	2;	$1+\tau$
$\frac{\vartheta_3}{\vartheta_4}$	1;	2τ

Let us observe how Jacobi made the transition from theta functions to $\operatorname{sn} u$, $\operatorname{cn} u$ and $\operatorname{dn} u$.

§4. Jacobi identities

Starting from the definition of theta functions, Jacobi derived numerous relations among them. These depend on the *fundamental identity*

$$\begin{aligned}&\vartheta_2(w)\vartheta_2(x)\vartheta_2(y)\vartheta_2(z)+\vartheta_3(w)\vartheta_3(x)\vartheta_3(y)\vartheta_3(z)\\&\quad=\vartheta_2(w')\vartheta_2(x')\vartheta_2(y')\vartheta_2(z')+\vartheta_3(w')\vartheta_3(x')\vartheta_3(y')\vartheta_3(z'),\end{aligned}\tag{I.4.1}$$

where the values w', x', y', z' are obtained from w, x, y, z by the linear transformation

$$\begin{aligned}2w' &= -w+x+y+z,\\2x' &= w-x+y+z,\\2y' &= w+x-y+z,\\2z' &= w+x+y-z.\end{aligned}\tag{I.4.2}$$

In order to derive this identity in the briefest way we note that ϑ_2 and ϑ_3 are special cases of the series

$$\Theta_{\mu,0}=\sum_{j=-\infty}^{\infty}\exp\left\{\pi i\left[(2j+\mu)x+\tau\frac{(2j+\mu)^2}{4}\right]\right\},$$

when $\mu=1$ and $\mu=0$. We rewrite the series in the form

$$\Theta_{\mu,0}(x)=\exp\left(-\frac{\pi i}{\tau}x^2\right)\sum_{j=-\infty}^{\infty}\exp\left\{\frac{\pi i}{\tau}\left(\frac{2j+\mu}{2}\tau+x\right)^2\right\}.\tag{I.4.3}$$

Letting $\mu=1$, successively substituting w, x, y, z for x, and equating the products of corresponding sides, we obtain a result which we write *symbolically* as

$$\vartheta_2(w)\vartheta_2(x)\vartheta_2(y)\vartheta_2(z)=\exp\left(-\frac{\pi i}{\tau}\sum w^2\right)\sum_{n,p,q,r}\exp\left[\frac{\pi i}{\tau}\sum\left(\frac{n}{2}\tau+w\right)^2\right].\tag{I.4.4}$$

For the sake of brevity we show only the first summand of each of the sums in the exponentials of powers. The rest of the summands are obtained by varying letters in alphabetical order (i.e., x, y, z come after w and p, q, r come after n). The summation indices n, p, q, and r run independently through all odd numbers.

Similarly, setting $\mu=0$ in (I.4.3) and using the same convention, we find

$$\vartheta_3(w)\vartheta_3(x)\vartheta_3(y)\vartheta_3(z)=\exp\left(-\frac{\pi i}{\tau}\sum w^2\right)\sum_{n,p,q,r}\exp\left[\frac{\pi i}{\tau}\sum\left(\frac{n\tau}{2}+w\right)^2\right],$$

where n, p, q, and r now run through all even numbers. Hence

$$\begin{aligned}&\vartheta_2(w)\vartheta_2(x)\vartheta_2(y)\vartheta_2(z)+\vartheta_3(w)\vartheta_3(x)\vartheta_3(y)\vartheta_3(z)\\&\quad=\exp\left(-\frac{\pi i}{\tau}\sum w^2\right)\sum_{n,p,q,r}\exp\left[\frac{\pi i}{\tau}\sum\left(\frac{n\tau}{2}+w\right)^2\right],\end{aligned}\tag{I.4.5}$$

where the summation is over all quadruples of integers n, p, q, and r of the same parity.

If the matrix

$$\begin{pmatrix} -\frac{1}{2} & \frac{1}{2} & \frac{1}{2} & \frac{1}{2} \\ \frac{1}{2} & -\frac{1}{2} & \frac{1}{2} & \frac{1}{2} \\ \frac{1}{2} & \frac{1}{2} & -\frac{1}{2} & \frac{1}{2} \\ \frac{1}{2} & \frac{1}{2} & \frac{1}{2} & -\frac{1}{2} \end{pmatrix}$$

is denoted by M, then the transformation given by (I.4.2) can be written concisely as

$$\begin{pmatrix} w' \\ x' \\ y' \\ z' \end{pmatrix} = M \begin{pmatrix} w \\ x \\ y \\ z \end{pmatrix}.$$

It is easy to check that conversely

$$\begin{pmatrix} w \\ x \\ y \\ z \end{pmatrix} = M \begin{pmatrix} w' \\ x' \\ y' \\ z' \end{pmatrix}$$

and moreover that the sum of the squares of the variables is invariant under the transformation (obviously orthogonal), i.e.,

$$\sum w'^2 = \sum w^2.$$

We substitute w', x', y', z' for w, x, y, z in (I.4.5) and (using the preceding equality) we first obtain

$$\begin{aligned} &\vartheta_2(w')\vartheta_2(x')\vartheta_2(y')\vartheta_2(z') + \vartheta_3(w')\vartheta_3(x')\vartheta_3(y')\vartheta_3(z') \\ &\quad = \exp\left(-\frac{\pi i}{\tau}\sum w'^2\right) \sum_{n,p,q,r} \exp\left[\frac{\pi i}{\tau}\sum\left(\frac{n\tau}{2} + w'\right)^2\right]. \qquad \text{(I.4.6)} \end{aligned}$$

We now apply the transformation (I.4.2) to the integral vectors

$$\begin{pmatrix} n \\ p \\ q \\ r \end{pmatrix}$$

with components of the same parity . It is easy to see that the set of these vectors is transformed into itself and the sum of the absolutely convergent series is not changed. Now the exponents under the summation sign take the form

$$\frac{\pi i}{\tau}\sum\left(\frac{n'\tau}{2} + w'\right)^2.$$

But if

$$\begin{pmatrix} n' \\ p' \\ q' \\ r' \end{pmatrix} = M \begin{pmatrix} n \\ p \\ q \\ r \end{pmatrix} \quad \text{and} \quad \begin{pmatrix} w' \\ x' \\ y' \\ z' \end{pmatrix} = M \begin{pmatrix} w \\ x \\ y \\ z \end{pmatrix},$$

then

$$\begin{pmatrix} n'\tau/2+w' \\ p'\tau/2+x' \\ q'\tau/2+y' \\ r'\tau/2+z' \end{pmatrix} = M \begin{pmatrix} n\tau/2+w \\ p\tau/2+x \\ q\tau/2+y \\ r\tau/2+z \end{pmatrix}$$

because the transformation is linear. Hence

$$\frac{\pi i}{\tau}\sum\left(\frac{n'\tau}{2}+w'\right)^2 = \frac{\pi i}{\tau}\sum\left(\frac{n\tau}{2}+w\right)^2,$$

and this means that the right-hand sides of equalities (I.4.5) and (I.4.6) are the same. Therefore, Jacobi's fundamental identity is proved.

Jacobi derives many corollaries which finally allow him to construct (with the help of theta functions) the solution of the problem of inversion of an elliptic integral. Let us follow his calculations.

First we add $1/2$ to w and to z, leaving x and y unchanged. Then, in view of (I.4.2), $1/2$ is added to x' and y' whereas w' and z' stay unchanged. Taking into consideration the first column of Table 1 and the identity (I.4.1) we obtain

$$\begin{aligned} &\vartheta_2(x)\vartheta_2(y)\vartheta_1(z)\vartheta_1(w) + \vartheta_3(x)\vartheta_3(y)\vartheta_4(z)\vartheta_4(w) \\ &\quad = \vartheta_1(x')\vartheta_1(y')\vartheta_2(z')\vartheta_2(w') + \vartheta_4(x')\vartheta_4(y')\vartheta_3(z')\vartheta_3(w'). \end{aligned} \tag{I.4.7}$$

Let $w = -z$, $y = x$. Then $w' = x+z$, $x' = 0$, $y' = 0$, $z' = x-z$. Therefore,

$$-\vartheta_2^2(x)\vartheta_1^2(z) + \vartheta_3^2(x)\vartheta_4^2(z) = \vartheta_4^2\vartheta_3(x-z)\vartheta_3(x+z). \tag{I.4.8}$$

Here and in what follows we use Jacobi's convention of omitting the argument of a theta function if it is zero (for example, $\vartheta_4 = \vartheta_4(0)$, $\vartheta_1 = \vartheta_1(0) = 0$). When $x = 0$ we obtain

$$\vartheta_2^2\vartheta_1^2(z) + \vartheta_4^2\vartheta_3^2(z) = \vartheta_3^2\vartheta_4^2(z). \tag{I.4.9}$$

Substituting $z + 1/2$ for z, we find that

$$\vartheta_2^2\vartheta_2^2(z) + \vartheta_4^2\vartheta_4^2(z) = \vartheta_3^2\vartheta_3^2(z). \tag{I.4.10}$$

In addition, substituting $z + (\tau/2)$ for z in (I.4.9) and checking with the second column of Table 1 we obtain another similar identity

$$-\vartheta_2^2M^2\vartheta_4^2(z) + \vartheta_4^2M^2\vartheta_2^2(z) = -\vartheta_3^2M^2\vartheta_1^2(z),$$

i.e.,

$$\vartheta_3^2\vartheta_1^2(z) + \vartheta_4^2\vartheta_2^2(z) = \vartheta_2^2\vartheta_4^2(z). \tag{I.4.11}$$

From (I.4.10) we obtain

$$\vartheta_2^4 + \vartheta_4^4 = \vartheta_3^4$$

when $z = 0$. Let

$$\sqrt{k} = \frac{\vartheta_2}{\vartheta_3}, \qquad \sqrt{k'} = \frac{\vartheta_4}{\vartheta_3}; \quad \text{then } k^2 + (k')^2 = 1. \tag{I.4.12}$$

We now introduce, using (I.4.11), a new variable φ by means of the formulas

$$\sin\varphi = \frac{\vartheta_3}{\vartheta_2}\frac{\vartheta_1(z)}{\vartheta_4(z)} = \frac{1}{\sqrt{k}}\frac{\vartheta_1(z)}{\vartheta_4(z)}, \qquad \cos\varphi = \frac{\vartheta_4}{\vartheta_2}\frac{\vartheta_2(z)}{\vartheta_4(z)} = \sqrt{\frac{k'}{k}}\frac{\vartheta_2(z)}{\vartheta_4(z)}. \qquad \text{(I.4.13)}$$

Finally, from (I.4.9) we derive

$$\frac{\vartheta_4}{\vartheta_3}\frac{\vartheta_3(z)}{\vartheta_4(z)} = \sqrt{1 - \frac{\vartheta_2^2\vartheta_1^2(z)}{\vartheta_3^2\vartheta_4^2(z)}} = \sqrt{1 - k^2\sin^2\varphi},$$

which can be written in the form

$$\Delta(\varphi) = \sqrt{k'}\,\frac{\vartheta_3(z)}{\vartheta_4(z)}. \qquad \text{(I.4.14)}$$

In order to conclude that the functions (I.4.13) and (I.4.14) coincide up to a scalar multiple of the argument with the functions $\operatorname{sn} x$, $\operatorname{cn} x$, and $\operatorname{dn} x$, Jacobi still had to derive addition theorems for the quotients $\alpha_i(z) = \vartheta_i(z)/\vartheta_4(z)$, $i = 1, 2, 3$, and then to find expressions for their derivatives. He did this as follows.

We return to the identity (I.4.7) and set $y' = z' = 0$, i.e., in view of (I.4.2), $w + x = 0$, $y - z = 0$. Then $w' = x + z$ and $x' = z - x$. We obtain

$$\begin{aligned} &-\vartheta_1(x)\vartheta_1(z)\vartheta_2(x)\vartheta_2(z) + \vartheta_3(x)\vartheta_3(z)\vartheta_4(x)\vartheta_4(z) \\ &\quad = \vartheta_3\vartheta_4\vartheta_3(x+z)\vartheta_4(x-z). \end{aligned}$$

On the other hand, it follows from (I.4.8) (with x replaced by $x + 1/2$) that

$$-\vartheta_1^2(x)\vartheta_1^2(z) + \vartheta_4^2(x)\vartheta_4^2(z) = \vartheta_4^2\vartheta_4(x+z)\vartheta_4(x-z).$$

Dividing the first result by the second, we obtain

$$\begin{aligned} &\frac{\vartheta_3}{\vartheta_4}\frac{\vartheta_3(x+z)}{\vartheta_4(x+z)} \\ &= \frac{(\vartheta_3(x)/\vartheta_4(x))\cdot(\vartheta_3(z)/\vartheta_4(z)) - (\vartheta_1(x)/\vartheta_4(x))(\vartheta_1(z)/\vartheta_4(z))\cdot(\vartheta_2(x)/\vartheta_4(x))(\vartheta_2(z)/\vartheta_4(z))}{1 - (\vartheta_1^2(x)/\vartheta_4^2(x))\cdot(\vartheta_1^2(z)/\vartheta_4^2(z))}, \end{aligned}$$

or more concisely

$$\alpha_3\cdot\alpha_3(x+z) = \frac{\alpha_3(x)\alpha_3(z) - \alpha_1(x)\alpha_1(z)\alpha_2(x)\alpha_2(z)}{1 - \alpha_1^2(x)\alpha_1^2(z)}. \qquad \text{(I.4.15)}$$

This is one of three *addition theorems for the ratios of the first three theta functions to the fourth.* The other two addition theorems are obtained by similar calculations and are essentially corollaries of the fundamental identity.

Formulas for derivatives follow easily from the addition theorems. Indeed, with help of (I.4.15) we find

$$\begin{aligned}\lim_{x\to 0}&\frac{\alpha_3[\alpha_3(z+x)-\alpha_3(z)]}{x}\\&=\lim_{x\to 0}\left\{\frac{\alpha_3(z)[\alpha_3(x)-\alpha_3]}{[1-\alpha_1^2(x)\alpha_1^2(z)]x}\right.\\&\qquad\left.-\frac{\alpha_1(z)\alpha_1(x)[\alpha_2(z)\alpha_2(x)-\alpha_1(z)\alpha_1(x)\alpha_3(z)\alpha_3]}{x[1-\alpha_1^2(x)\alpha_1^2(z)]}\right\}\\&=-\alpha_1(z)\alpha_1'\alpha_2(z)\alpha_2,\end{aligned}$$

where we used the fact that $\alpha_1 = \alpha_1(0) = 0$, and also $\alpha_3' = \alpha_3'(0) = 0$ because $\alpha_3(x)$ is even. Therefore

$$\alpha_3\alpha_3'(z) = -\alpha_1'\alpha_2\alpha_1(z)\alpha_2(z),$$

i.e.,

$$\frac{d}{dz}\left[\frac{\vartheta_3(z)}{\vartheta_4(z)}\right] = -\frac{\vartheta_2\vartheta_1'}{\vartheta_4\vartheta_3}\frac{\vartheta_1(z)\vartheta_2(z)}{\vartheta_4^2(z)}. \tag{I.4.16}$$

Similarly, we find the other two derivatives from the corresponding addition theorems:

$$\frac{d}{dz}\left[\frac{\vartheta_1(z)}{\vartheta_4(z)}\right] = \frac{\vartheta_4\vartheta_1'}{\vartheta_2\vartheta_3}\frac{\vartheta_2(z)\vartheta_3(z)}{\vartheta_4^2(z)}, \qquad \frac{d}{dz}\left[\frac{\vartheta_2(z)}{\vartheta_4(z)}\right] = -\frac{\vartheta_3\vartheta_1'}{\vartheta_4\vartheta_2}\frac{\vartheta_1(z)\vartheta_3(z)}{\vartheta_4^2(z)}. \tag{I.4.17}$$

Differentiating (I.4.14) and using (1.4.13) we obtain

$$\begin{aligned}\sqrt{k'}\frac{d}{dz}\left[\frac{\vartheta_3(z)}{\vartheta_4(z)}\right] &= -k^2\frac{\sin\varphi\cos\varphi}{\Delta(\varphi)}\frac{d\varphi}{dz}\\&= -k\sqrt{k'}\,\frac{\vartheta_1(z)\vartheta_2(z)}{\vartheta_4^2(z)}\frac{1}{\Delta(\varphi)}\frac{d\varphi}{dz},\end{aligned}$$

or, in view of (I.4.16) and (I.4.12),

$$\frac{\vartheta_3\vartheta_1'}{\vartheta_2\vartheta_4}\cdot\frac{dz}{d\varphi} = \frac{1}{\Delta(\varphi)}.$$

Jacobi finds (using extremely clever computations) the following expression for ϑ_1:

$$\vartheta_1' = \pi\vartheta_2\vartheta_3\vartheta_4.$$

Hence the preceding formula can be expressed in the form

$$\pi\vartheta_3^2\frac{dz}{d\varphi} = \frac{1}{\sqrt{1-k^2\sin^2\varphi}}. \tag{I.4.18}$$

We note that Jacobi is interested primarily in real values of functions. Hence he considers the parameter $q = e^{\pi i\tau}$ in the expression for theta functions to be real (to guarantee this, it is sufficient to require τ to be pure imaginary). Since then ϑ_2, ϑ_3, and ϑ_4 are also real (positive), in view of (I.3.4), (I.3.5), and (I.3.6), k and k' defined by the formulas (I.4.12) are real and positive, and moreover less than 1. It follows from (I.4.18) that when φ is real, z increases with φ. Since it can be seen from the formula (I.4.13) defining $\sin\varphi$ that $\sin\varphi = 0$ when $z = 0$ and $\sin\varphi = 1$ when $z = 1/2$ (because $\vartheta_1(1/2)/\sqrt{k}\vartheta_4(1/2) = \vartheta_2/\sqrt{k}\vartheta_3 = 1$), we find when we integrate (I.4.18) from 0 to $\pi/2$ that

$$\frac{\pi}{2}\vartheta_3^2 = \int_0^{\pi/2} \frac{d\varphi}{\sqrt{1-k^2\sin^2\varphi}} = K. \tag{I.4.19}$$

Set $2Kz = u$; then we find from (I.4.18) that

$$u = \int_0^{\varphi} \frac{d\varphi}{\sqrt{1-k^2\sin^2\varphi}}, \qquad \text{i.e.}, \ \varphi = \operatorname{am} u.$$

Finally, formulas (I.4.13) and (I.4.14) give

$$\begin{aligned} \sin\operatorname{am} u &= \operatorname{sn} u = \frac{1}{\sqrt{k}}\frac{\vartheta_1(u/2K)}{\vartheta_4(u/2K)}, \\ \cos\operatorname{am} u &= \operatorname{cn} u = \sqrt{\frac{k'}{k}}\frac{\vartheta_2(u/2K)}{\vartheta_4(u/2K)}, \\ \Delta\operatorname{am} u &= \operatorname{dn} u = \sqrt{k'}\frac{\vartheta_3(u/2K)}{\vartheta_4(u/2K)}. \end{aligned} \tag{I.4.20}$$

It follows by comparing formulas (I.4.20) with formulas from "Fundamenta Nova" given at the beginning of §3 that

$$\mathrm{H}(u) = \vartheta_1\left(\frac{u}{2K}\right), \qquad \Theta(u) = \vartheta_4\left(\frac{u}{2K}\right).$$

Thus the task Jacobi set for himself: *starting from a priori constructed theta functions, obtain a solution of the problem of inversion of an elliptic integral* is accomplished. To be sure, he also occupied himself with the question of whether it is always possible to chose a parameter q such that an arbitrarily prescribed value for the modulus k $(0 < k < 1)$ is obtainable. However, we will not discuss this subject here.

§5. Problem of inversion of a hyperelliptic integral

A brief memoir by Jacobi "On quadruply periodic functions of two variables, which arise in the theory of Abelian functions", appeared in Latin in Crelle's journal in 1835 and was published in German with comments by H. Weber in the series of Ostwald's classics (number 64). Already in "Fundamenta Nova" Jacobi stated his conviction that double periodicity is, so to speak, the highest periodicity that can be encountered in analysis (for a function of one variable). The first part of the memoir cited above is devoted to justifying this assertion. The arguments are based on a general analysis of the property of periodicity of a function. We do not need to discuss this part of Jacobi's memoir here, since the next chapter will be devoted to a modern treatment of the problem for a function of any number of variables. We note only that Jacobi came to the following conclusion by a purely arithmetic argument: *If a given function has three periods, then either they are constructible from two* (in a form of an integral linear combination) *or the function must have a period smaller than any given value.* "Since this is absurd", concludes Jacobi, "a triply periodic function does not exist."

We will return later to this statement, which played an important role in the birth of the theory of Abelian functions. For the present, we note only that Jacobi's assertion is hard to contest because, quite in the spirit of his time, he did not specify the class of functions under consideration. For single-valued analytic functions of one complex variable, Jacobi's conclusion is of course valid. Indeed, if we assume that $f(z) \neq \text{const}$ and that there exists a sequence of nonzero periods $\{\omega_n\}$ converging to zero, then we immediately conclude that, because the functions are single-valued, the relations $f(z_0 + \omega_n) = f(z_0)$, $n = 1, 2, \dots,$ cannot hold. This interpretation was accepted by C. Hermite and H. Weber. But the point is that Jacobi, as we will show later, did not restrict himself to the consideration of single-valued or even finite-valued functions. Without such a restriction the preceding argument is not applicable. Moreover, he never refers to the fact that a function is representable by a power series (is analytic). Incidentally, when Jacobi speaks of an analytic function he means: *a function encountered or used in analysis.* In the latter case a very simple example shows that the presence of infinitely small periods of a continuous single-valued function does not lead to a contradiction. Thus, letting $f(x + iy) = \sin x$ we get a function which besides real periods (multiples of 2π) also has pure imaginary periods $i\eta$, where η is any real number and in particular can be arbitrarily small in absolute value. However, we will return to questions of this kind later.

Discarding triply periodic function as nonexistent, Jacobi stressed the fact that the argument does not rule out the existence of functions of several variables with more than two periods and refers to his note published in the same journal three years earlier. He was planning to clarify this important point again later. All his arguments concern the hyperelliptic integral of the first kind

$$u = \int_0^x \frac{(\alpha + \beta x)\, dx}{\sqrt{\mathfrak{X}}}, \tag{I.5.1}$$

where $\mathfrak{X}$ is a quintic polynomial whose pairwise distinct zeros lie on the real axis. Jacobi lets

$$\mathfrak{X} = x(1 - x)(1 - \varkappa^2 x)(1 - \lambda^2 x)(1 - \mu^2 x), \tag{I.5.2}$$

where $1 > \varkappa^2 > \lambda^2 > \mu^2 > 0$.

We say beforehand that the discussion will be about an attempt to invert the integral (I.5.1), i.e., to consider a function $x = \lambda(u)$ by analogy with the inversion of the elliptic integral of the first kind. Here Jacobi's goal is to establish the impossibility of such an inversion, based on the result that the function $\lambda(u)$ would have four periods if it existed. He had just declared such functions to be meaningless.

Let us follow his argument, with simplifications that do not affect the essence of the matter. He uses an interesting method of determining the periods of the integrals (I.5.1) without using the complex plane (Jacobi never uses complex integration, which had been introduced by Cauchy ten years earlier). This method is based on a substitution of the form $x = R(\sin^2\varphi)$, where R is a rational function selected in such a way that when φ increases from 0 to $\pi/2$ the variable x also increases and runs through one of the six intervals into which the zeros of $\mathfrak{X}$ divide the real axis. For each interval its own function R is selected. For example, for the interval $(-\infty, 0)$ it might be the substitution $x = 1 - 1/\sin^2\varphi$, for the interval (0,1) it might be $x = \sin^2\varphi$, and so on.

Let us denote, for brevity, the integrand by $f(x)$ and the corresponding interval by (a, b). We notice that $f(x)$ has branch points of the first order at which it takes the value ∞. These branch points are the endpoints of the interval (zeros of $\mathfrak{X}$). After the substitution the integrand becomes $F(\varphi) = f(x)\,dx/d\varphi$ and the endpoints 0, $\pi/2$ corresponding to a and b are no longer singular points of the integrand. Thus, the integral $u = \int_0^x f(x)\,dx$ takes the form

$$u = A + \int_0^\varphi F(\varphi)\,d\varphi,$$

where $A = \int_0^a f(x)\,dx$, and the function $F(\varphi)$ has the following properties: (1) it has a period equal to π, (2) it is even, and (3) it is analytic on the segment $[0, \pi]$ (as we already mentioned above, Jacobi never stipulates the last property).

Because of these properties, $F(\varphi)$ can be represented as a uniformly convergent Fourier series of the form

$$F(\varphi) = \gamma_0 + \gamma_1 \cos 2\varphi + \gamma_2 \cos 4\varphi + \cdots.$$

Using this series and integrating termwise, we find

$$u = A + \gamma_0\varphi + \frac{1}{2}\gamma_1 \sin 2\varphi + \frac{1}{4}\gamma_2 \sin 4\varphi + \cdots. \tag{I.5.3}$$

The substitution of $\varphi + \pi$ for φ does not change the value of $x = R(\sin^2\varphi)$ and increases u by $\pi\gamma_0$. Using this, Jacobi then concludes that if x is considered to be a function of u, it must have a period $\pi\gamma_0$. Set $\varphi = \pi/2$ in (I.5.3). We get

$$A + \int_0^{\pi/2} F(\varphi)\,d\varphi = A + \int_a^b f(x)\,dx = A + \frac{\pi}{2}\gamma_0.$$

Therefore

$$2\int_a^b f(x)\,dx = \pi\gamma_0 \tag{I.5.4}$$

is a period of the function $x = \lambda(u)$.

Corresponding to six intervals $(-\infty, 0)$, $(0, 1)$, $(1, 1/\varkappa^2)$, $(1/\varkappa^2, 1/\lambda^2)$, $(1/\lambda^2, 1/\mu^2)$ and $(1/\mu^2, +\infty)$ Jacobi first finds six periods of the inverse of the hyperelliptic integral:

$$2\int_{-\infty}^{0} = 2u_1\sqrt{-1}, \qquad 2\int_0^1 = 2u_2, \qquad 2\int_1^{1/\varkappa^2} = 2u_3\sqrt{-1},$$

$$2\int_{1/\varkappa^2}^{1/\lambda^2} = 2u_4, \qquad \int_{1/\lambda^2}^{1/\mu^2} = 2u_5\sqrt{-1}, \qquad 2\int_{1/\mu^2}^{\infty} = 2u_6. \tag{I.5.5}$$

Three of these are pure imaginary (first, third, and fifth) and the other three are real. The values of the roots under the integral sign are selected in such a way that the numbers u_j, $j = 1, 2, \dots, 6$, are positive. Jacobi's mathematical intuition allowed him to see that the integral

$$2\int_{-\infty}^{+\infty} \frac{\alpha + \beta x}{\sqrt{\mathfrak{X}}}\,dx$$

could not be obtained by simply adding the numbers (I.5.5). Today we would say that the choice of one of the two branches of the function $(\alpha + \beta x)/\sqrt{\mathfrak{X}}$ on the upper half plane and on its boundary is completely determined by the fact that the positive value of the radical is taken over the interval (0,1). In any case Jacobi, in complete agreement with this choice, wrote

$$-u_1\sqrt{-1} + u_2 + u_3\sqrt{-1} - u_4 - u_5\sqrt{-1} + u_6 = \int_{-\infty}^{+\infty} \frac{\alpha + \beta x}{\sqrt{\mathfrak{X}}}\,dx.$$

And the last integral, he continued, is equal to zero since both limits become zero under the substitution of $1/x$ for x.

Such an argument does not prove anything, of course. However, the integral $\int_{-\infty}^{+\infty}$ is indeed equal to zero, as follows from Cauchy's integral theorem (as usual, we go around the branch points, including ∞, on semicircles and then take the limit). As a result, two relations for the periods (I.5.5) are obtained. These are

$$u_1 + u_5 = u_3, \qquad u_2 + u_6 = u_4. \tag{I.5.6}$$

It follows from these that there are four periods instead of six:

$$2u_2 = 2\int_0^1 \frac{\alpha + \beta x}{\sqrt{\mathfrak{X}}}\,dx, \qquad 2u_6 = 2\int_{1/\mu^2}^{\infty} \frac{\alpha + \beta x}{\sqrt{\mathfrak{X}}}\,dx,$$

$$2u_1\sqrt{-1} = 2\sqrt{-1}\int_{-\infty}^{0} \frac{\alpha + \beta x}{\sqrt{-\mathfrak{X}}}\,dx, \tag{I.5.7}$$

$$2u_5\sqrt{-1} = 2\sqrt{-1}\int_{1/\lambda^2}^{1/\mu^2} \frac{\alpha + \beta x}{\sqrt{-\mathfrak{X}}}\,dx$$

(the value of the radicals under the integral sign are chosen positive).

Thus, the function $x = \lambda(u)$, inverse to the integral (I.5.1), has four periods and therefore, as was proved by Jacobi in the first part of this memoir, must have infinitely small periods. Jacobi continues "*Hence the function does not change when it then takes all real or imaginary values. Otherwise among the values it can then take when $x = \lambda(u)$ is fixed there must always be such that differ from an arbitrary real or imaginary number by a quantity less than any given value. But this is absurd.*"

We have already mentioned that Jacobi does not at all discard infinite-valued functions. On the contrary, he mentions logarithms, inverse trigonometric functions, elliptic integrals, and so on. However, in all these cases, which today we think of as well-known, the branch points are isolated in the plane and it is not at all hard to define single-valued branches over any simply connected region containing no branch points. But here , as we will now see, branch points (more precisely, their projections on the complex plane) are everywhere dense and Jacobi does not see any possibility of separating single-valued branches. Precisely this circumstance forces him to call such a situation absurd (obviously from the point of view of all previous mathematical experience).

Jacobi's difficulties can be easily cleared up from the point of view of interpreting analytic functions as mappings from one region to another. This point of view was absolutely foreign to Jacobi. It was first introduced into mathematics by Riemann in his fundamental dissertation in 1851. Let us consider the mapping (I.5.1) from a modern point of view. Obviously this hyperelliptic integral is an integral of Schwarz-Christoffel type. If we select a single-valued branch of the function $(\alpha + \beta z)/\sqrt{\mathfrak{X}(x)}$ (α and β are real numbers) over the upper half plane $\operatorname{Im} z > 0$ $(z = x + iy)$ by selecting positive values of the radical for $0 < z = x < 1$, then the mapping (I.5.1) maps the real axis to the boundary of a hexagon. Jacobi found the lengths of the sides of this hexagon but he did not represent them geometrically, as in Figure 3 (p. 22), for example.

The vertices O, A, B, C, D, and E are images of the points 0, 1, $1/\varkappa^2$, $1/\lambda^2$, $1/\mu^2$, ∞. The validity of the relations (I.5.6) follows immediately from the form of this hexagon.

The function $x = \lambda(u)$ (inverse to the integral (I.5.1)), or more precisely a single-valued branch of this function, is immediately defined inside $OABCDE$ and can be extended to the whole plane by analytic continuation (Schwarz reflection principle). Its periods are found to be $2u_2$, $2u_6$, $2u_1 i$, $2u_5 i$.

It is obvious that for the special values of $\varkappa^2$, λ^2, μ^2, α, and β on which the integral (I.5.1) depends, the quotient u_6/u_2 or the quotient u_5/u_1, and possibly both, can be rational. Then the number of basic periods becomes three or even two. Hence quadruple periodicity assumes the absence of

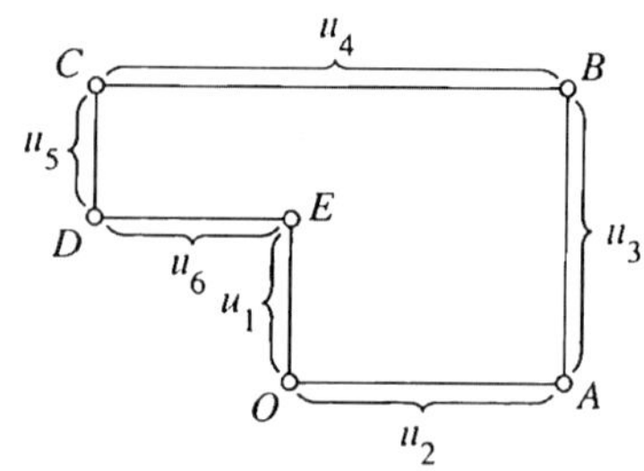

FIGURE 3

special relations (expressed by the condition that u_6/u_2 or u_5/u_1 is rational) between the parameters mentioned above. We will exclude such cases and will suppose both quotients to be irrational.

Notice that, given a particular hexagon of the type of Figure 3, we can always find (using the Schwarz-Christoffel theorem) parameters such that the half plane is mapped onto a polygon with the same ratios of sides as the given polygon. Then $\lambda(u)$ is indeed *quadruply periodic*. Thus, reflecting $OABCDE$ twice, first across the side BC and then (the resulting figure) across the side $D'E'$, we convince ourselves by looking at the graph that $2u_5 i$ is indeed a period of our function. See Figure 4.

If we continue the function analytically everywhere by reflecting hexagons across their sides, they form an infinite-sheeted Riemann surface with branch points of the first order at the vertices. The projections of the branch points onto the complex plane are everywhere dense.

However, a single-valued, and indeed one-sheeted, branch $\lambda(u)$ is defined on each hexagon. Then $\lambda(u)$ maps the hexagon conformally on the upper or lower half-plane, depending on whether this polygon is obtained from the original one by an even or odd number of reflections. Since periods of $\lambda(u)$ are numbers of the form

$$mu_2 + nu_4 + i(pu_1 + qu_5),$$

where m, n, p, and q are arbitrary integers, we conclude, in view of our

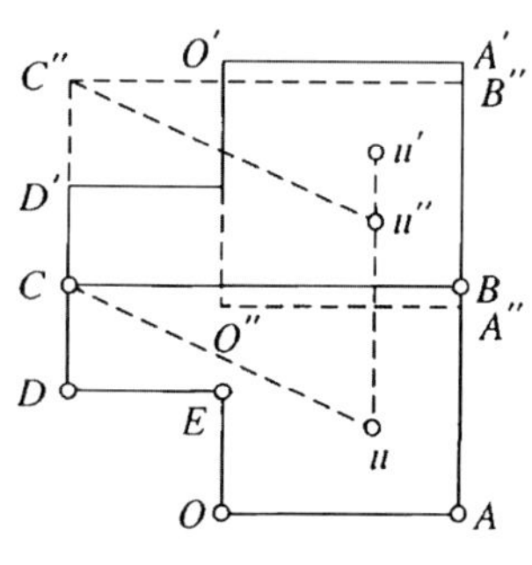

FIGURE 4

assumption that u_6/u_2 and u_5/u_1 are irrational, that any point on the plane is a limit point of the periods and therefore infinitely small periods exist, as was asserted by Jacobi. But points infinitely close on the plane and at which $\lambda(u)$ assumes the same value x belong to *different sheets of the Riemann surface*, i.e., they are quite far from each other on that surface. Notice here that we cannot just use continuous continuation in place of analytic continuation. This fact is, of course, quite unusual from the point of view of analysis in the days before Riemann.

We repeat that this entire way of thinking was foreign to Jacobi. The reason was, of course, not that the Schwarz reflection principle had not yet been stated, because its application to the special case of the hyperelliptic integral (I.5.1) does not necessarily assume the general theory. The point is that the concept of a *function as a mapping of a region onto a region* would have been necessary. And this concept was lacking.

§6. Problem of Jacobi. Göpel and Rosenhain

If asserting the very impossibility of inverting the hyperelliptic integral (I.5.1) can be called a false lead of Jacobi's, this false lead turned out to be quite fruitful. It forced him to search for a way out by stating the problem in a totally new form: namely, to consider the upper limit of the integral as a function not of one but of at least two variables. He thought, totally justifiably, that for a function of two variables the assumption about the presence of four periods does not lead to the existence of infinitely small periods. We can assume as a hypothesis that he actually constructed quadruply periodic single-valued functions of two variables starting from an attempt to generalize the notion of theta function to the case of two variables and then to consider their quotients as periodic functions. In any case the path of research by Göpel (1847) and Rosenhain (1851), who followed in Jacobi's footsteps, was exactly this.

Postponing this discussion until the end of this section, we assume together with Jacobi that $\lambda(u, u')$ is a periodic function of two complex variables. What should a period of such a function mean? Simultaneous consideration of two hyperelliptic integrals (we will discuss this later), or perhaps the experience of generalizing theta functions, led Jacobi to the idea that a period must no longer be defined as a single number as in the case of a function of one variable, but as an ordered pair of complex numbers ω and ω' such that

$$\lambda(u+\omega, u'+\omega') = \lambda(u, u'),$$

no matter what u and u' are taken. If the function $\lambda(u, u')$ is quadruply periodic then this must mean that it has four independent periods (two-dimensional complex vectors $\Omega^{(k)} = \binom{\omega_k}{\omega'_k}$, $k = 1, 2, 3, 4$; where ω_k, ω'_k are complex numbers), i.e., such that from the relation

$$m_1\Omega^{(1)} + m_2\Omega^{(2)} + m_3\Omega^{(3)} + m_4\Omega^{(4)} = 0$$

(where m_1, m_2, m_3, m_4 are integers) it necessarily follows that

$$m_1 = m_2 = m_3 = m_4 = 0.$$

Why doesn't the same conclusion about the presence of infinitely small periods have to hold now? Because (as Jacobi answers) if we consider a sequence of quadruples of integers $(m_1^{(j)}, m_2^{(j)}, m_3^{(j)}, m_4^{(j)})$ not all simultaneously zero and such that for example

$$\lim_{j\to\infty} (m_1^{(j)}\omega_1 + m_2^{(j)}\omega_2 + m_3^{(j)}\omega_3 + m_4^{(j)}\omega_4) = 0,$$

i.e., if the first components of the sequence of vectors converge to zero, then the second components $m_1^{(j)}\omega_1' + m_2^{(j)}\omega_2' + m_3^{(j)}\omega_3' + m_4^{(j)}\omega_4'$ converge (in the case of the hyperelliptic integral he considered) to ∞. And this means that there are no infinitely small periods generated by integral combinations of Ω^k $(k = 1, 2, 3, 4)$.

Jacobi does not give a proof of this assertion. The essence of the mechanism that causes such a phenomenon is easy to understand by using an elementary example. Consider the expressions $m + n\sqrt{2}$ and $m + n\sqrt{3}$, where m and n are integers, not simultaneously zero. If $m^{(j)} + n^{(j)}\sqrt{2} \to 0$, then $n^{(j)}(m^{(j)}/n^{(j)} + \sqrt{2}) \to 0$, i.e., $m^{(j)}/n^{(j)} \to -\sqrt{2}$ and $n^{(j)} \to \infty$. Hence

$$m^{(j)} + n^{(j)}\sqrt{3} = n^{(j)}\left(\frac{m^{(j)}}{n^{(j)}} + \sqrt{3}\right) \approx n^{(j)}(\sqrt{3} - \sqrt{2}) \to \infty.$$

Incidentally, the absence of infinitely small periods can easily be proved under the additional assumption that the vectors $\Omega^{(k)}$ are linearly independent over the reals, i.e., it follows from the equality

$$\sum_{k=1}^{4} \alpha_k \Omega^{(k)} = 0, \tag{I.6.1}$$

where α_k are real numbers, that $\alpha_k = 0$, $k = 1, 2, 3, 4$. Indeed, assume that

$$\lim_{j\to\infty} \sum_{k=1}^{4} m_k^{(j)} \Omega^{(k)} = 0, \tag{I.6.2}$$

where $m_k^{(j)}$ are integers such that $\sum_{k=1}^4 m_k^{(j)} = M_j^2 \neq 0$. Then the sequence $\{M_j\}$ must be unbounded (otherwise the sequence (I.6.2) would contain an infinite set of terms equal to each other and therefore be equal to zero, contrary to the condition of linear independence). Restricting ourselves to a subsequence we can assume that $\lim_{j\to\infty} M_j = \infty$. Now divide all terms of the sequence (I.6.2) by M_j. Then we obtain a new sequence $\{\sum_{k=1}^4 \mu_k^{(j)}\Omega^{(k)}\}$ converging to zero, where $\mu_k^{(j)}$ are real numbers such that $\sum_{k=1}^4 (\mu_k^{(j)})^2 = 1$. Again restricting ourselves to a subsequence such that the limits $\lim_{j\to\infty} \mu_k^{(j)} = \mu_k$, $k = 1, 2, 3, 4$, exist, and taking the limit, we find that $\sum_{k=1}^{4} \mu_k \Omega_k = 0$, where $\sum_{k=1}^{4} \mu_k^2 = 1$. But this contradicts the assumption about the linear independence of the periods over the reals.

It is clear now why the idea of "distributing" four periods of a hyperelliptic integral between two variables occurred to Jacobi.

With this purpose in mind, in 1832 he considers (motivated by considerations connected with Abel's theorem (see section I.8)) two hyperelliptic integrals of the first kind $\int (\alpha + \beta x)\, dx/\sqrt{\mathfrak{X}}$ and $\int (\alpha' + \beta' x)\, dx/\sqrt{\mathfrak{X}} (\alpha' \beta - \alpha \beta' \neq 0)$ simultaneously and forms their sums

$$\int_a^x \frac{\alpha + \beta x}{\sqrt{\mathfrak{X}}}\, dx + \int_b^y \frac{\alpha + \beta x}{\sqrt{\mathfrak{X}}}\, dx = u,$$
$$\int_a^x \frac{\alpha' + \beta' x}{\sqrt{\mathfrak{X}}}\, dx + \int_b^y \frac{\alpha' + \beta' x}{\sqrt{\mathfrak{X}}}\, dx = u'.$$

The new inversion problem that he states consists of treating the limits x and y of both integrals as *functions of two complex variables* u and u':

$$x = \lambda(u, u'), \qquad y = \lambda'(u, u').$$

These functions also turn out to be quadruply periodic. This is what Jacobi proves in his memoir of 1835 on quadruply periodic funtions.

Since the polynomial $\mathfrak{X}$ remains the same, in order to determine the periods by Jacobi's method we can again use a trigonometric substitution $x = R(\sin^2 \varphi)$ (see section I.5) for each interval between adjacent zeros of $\mathfrak{X}$. Using the formula (I.5.3) we obtain

$$\int_a^x \frac{\alpha + \beta x}{\sqrt{\mathfrak{X}}}\, dx = A + \gamma_0 \varphi + \frac{1}{2}\gamma_1 \sin 2\varphi + \cdots, \tag{I.6.3}$$

$$\int_a^x \frac{\alpha' + \beta' x}{\sqrt{\mathfrak{X}}}\, dx = A' + \gamma_0' \varphi + \frac{1}{2}\gamma_1' \sin 2\varphi + \cdots \tag{I.6.4}$$

(here our notation differs slightly from that of Jacobi). Similarly, for the substitution $y = R_1(\sin^2 \psi)$, where y runs through another (or the same) interval between two zeros of $\mathfrak{X}$, we obtain

$$\int_b^y \frac{\alpha + \beta x}{\sqrt{\mathfrak{X}}}\, dx = B + \delta_0 \psi + \frac{1}{2}\delta_1 \sin 2\psi + \cdots, \tag{I.6.5}$$

$$\int_b^y \frac{\alpha' + \beta' x}{\sqrt{\mathfrak{X}}}\, dx = B' + \delta_0' \psi + \frac{1}{2}\delta_1' \sin 2\psi + \cdots. \tag{I.6.6}$$

If φ is now increased by $m\pi$ and ψ is increased by $m'\pi$, where m and m' are arbitrary integers, then x and y do not change; but u, which is the sum of integrals (I.6.3) and (I.6.5), changes by $m\pi\gamma_0 + m'\pi\delta_0$, and u' changes by $m\pi\gamma_0' + m'\pi\delta_0'$. If we replace $\pi\gamma_0$, $\pi\delta_0$, $\pi\gamma_0'$ and $\pi\delta_0'$ by their expressions (I.5.4) as twice the integrals taken over the same intervals as before, and recall their expressions in the form (I.5.6), we arrive at the following statement, which Jacobi calls fundamental.

Fundamental theorem (Jacobi). *Let*

$$\mathfrak{X} = x(1-x)(1-\varkappa^2 x)(1-\lambda^2 x)(1-\mu^2 x),$$

$$2\int_{-\infty}^{0} \frac{\alpha+\beta x}{\sqrt{-\mathfrak{X}}}\,dx = i_1, \qquad 2\int_{0}^{1} \frac{\alpha+\beta x}{\sqrt{\mathfrak{X}}}\,dx = i_2,$$

$$2\int_{1/\lambda^2}^{1/\mu^2} \frac{\alpha+\beta x}{\sqrt{-\mathfrak{X}}}\,dx = i_3, \qquad 2\int_{1/\mu^2}^{\infty} \frac{\alpha+\beta x}{\sqrt{\mathfrak{X}}}\,dx = i_4,$$

$$2\int_{-\infty}^{0} \frac{\alpha'+\beta' x}{\sqrt{-\mathfrak{X}}}\,dx = i_1', \qquad 2\int_{0}^{1} \frac{\alpha'+\beta' x}{\sqrt{\mathfrak{X}}}\,dx = i_2',$$

$$2\int_{1/\lambda^2}^{1/\mu^2} \frac{\alpha'+\beta' x}{\sqrt{-\mathfrak{X}}}\,dx = i_3', \qquad 2\int_{1/\mu^2}^{\infty} \frac{\alpha'+\beta' x}{\sqrt{\mathfrak{X}}}\,dx = i_4';$$

take $x = \lambda(u, u')$, $y = \lambda'(u, u')$ *where*

$$\int_a^x \frac{\alpha+\beta x}{\sqrt{\mathfrak{X}}}\,dx + \int_b^y \frac{\alpha+\beta x}{\sqrt{\mathfrak{X}}}\,dx = u,$$

$$\int_a^x \frac{\alpha'+\beta' x}{\sqrt{\mathfrak{X}}}\,dx + \int_b^y \frac{\alpha'+\beta' x}{\sqrt{\mathfrak{X}}}\,dx = u'.$$

Then

$$\begin{aligned}\lambda(u, u') = \lambda(&u + m i_1\sqrt{-1} + m' i_2 + m'' i_3\sqrt{-1} + m''' i_4;\\ &u' + m i_1'\sqrt{-1} + m' i_2' + m'' i_3'\sqrt{-1} + m''' i_4'),\\ \lambda'(u, u') = \lambda'(&u + m i_1\sqrt{-1} + m' i_2 + m'' i_3\sqrt{-1} + m''' i_4;\\ &u' + m i_1'\sqrt{-1} + m' i_2' + m'' i_3'\sqrt{-1} + m''' i_4').\end{aligned}$$

Jacobi does not assert that λ and λ' are single-valued functions. Notice that if $a = b$ it is obvious that u and u' do not change when x and y are interchanged. Hence it can be expected that $x + y$ and xy will be single-valued. Consequently Jacobi proposes to consider x and y as the roots of a quadratic equation $Uz^2 - U'z + U'' = 0$. Apparently he is convinced that the functions $U = U(u, u')$, $U' = U'(u, u')$, $U'' = U''(u, u')$ are single-valued (which is indeed the case). These functions (single-valued and quadruply periodic), and their generalizations, are *the object of study of the subsequent theory of Abelian functions*.

Jacobi himself did not construct a general theory of quadruply periodic (or, more generally, multiply periodic) functions. As for the problem he proposed about the study of limits of two hyperelliptic integrals of the first kind as functions of two sums of such integrals, solutions were given by two German mathematicians: Göpel (1847) and Rosenhain (published after a five year delay in 1851). They reached their results independently of each other but their results were entirely consistent with the spirit of the ideas and the results of Jacobi himself.

Both of these mathematicians first build the theory of theta functions of two independent variables u and u'. In this case, instead of four functions $\theta_j(u)$, $j = 1, 2, 3, 4$, considered by Jacobi in one variable, there are 16 functions. These functions (being entire) have a property of generalized periodicity. Namely: when u (or u') changes by a nonzero value ω (or ω'), they gain factors which have the form of an exponential function with exponents which are linear with respect to u and u' (See Table 1 in section I.3. The table deals with the case of one variable; we shall discuss the general theory of theta functions of many variables in Chapter IV.) Quadruply periodic single-valued functions of two variables were then defined as appropriately constructed quotients of theta functions.

From the method of the construction itself, it follows that these are meromorphic functions of u and u' (i.e., functions representable as quotients of two entire functions (see section II.2)). By the way, the notion of the class of meromorphic functions was itself formulated somewhat later (Briot and Bouquet, 1856). It was proved (after quite clever but still rather complicated calculations) that with the help of such functions it is indeed possible to solve Jacobi's problem in the case of hyperelliptic integrals corresponding to a polynomial of fifth or sixth degree. Actually, the flow of the argument in the general investigation was not the same as could have been assumed on the basis of what we have just said. It is essential to stress that Rosenhain, who was a student of Jacobi, followed Jacobi quite closely in his work about inversion (1832, 1835). The starting point is indeed the inversion problem in the form explained above and theta functions are introduced as an instrument for its solution. Göpel, on the other hand, begins with a general consideration of theta functions, investigates algebraic relations between them, differentiation formulas, etc., introduces quotients of theta functions, and determines differential equations for these quotients. Only at the end of his work does he show that the functions studied by him allow one to solve Jacobi's inversion problem. In its basic and important features the results of Göpel can be likened to Jacobi's inversion method translated to the case of two variables, whereas the results of Rosenhain are closer to the path originally taken by Jacobi in his "New Foundations".

§7. Algebraic functions and their Riemann surfaces

In this section and the next, we will give an overview of facts which will enable us to state Jacobi's inversion problem in its most general form. We do not, however, use these facts in further developments, and reasons for this will be given at the end of this section.

Here are the main notions and facts of the theory of algebraic functions, in a form close to V. Puiseux's memoir of 1850, with the one significant difference that we will consider algebraic functions and their integrals (Abelian integrals) on Riemann surfaces. The notion of Riemann surface, as is well-known, was first introduced by Riemann in his famous dissertation "Foundations of the general theory of functions of a complex variable" (1851).

Let $f(\overset{m}{z}, \overset{n}{w})$ be a polynomial with complex coefficients (the numbers above z and w show, as was suggested by Riemann, the degrees of the polynomial with respect to z and w). We will assume that the polynomial is *irreducible* over the field of complex numbers, i.e., that it cannot be expressed as a product of polynomials of lower degree (none of which is a constant). Write the polynomial in the form

$$f(z, w) = A_0(z)w^n + \cdots + A_n(z), \tag{I.7.1}$$

where $A_j(z)$ $(j = 0, \ldots, n)$ are polynomials of degree no larger than m with no zeros in common, and moreover $A_0(z) \not\equiv 0$, $A_n(z) \not\equiv 0$, $n \geq 1$.

Form the resultant of the polynomials f and $\partial f/\partial w$ and denote it by $R_w(z)$. As we know from algebra, $R_w(z)$ is the polynomial

$$R_w(z) = n^2 A_0^{2n-1}(z) \prod_{i=1}^{n} \prod_{j=1}^{n-1} (\alpha_i - \beta_j), \tag{I.7.2}$$

where $\alpha_i = \alpha_i(z)$ are the roots of the equation $f(z, w) = 0$ and $\beta_j = \beta_j(z)$ are the roots of the equation $f'_w(z, w) = 0$ (here we are talking about values of w for an arbitrary fixed z). The zeros of the resultant exhaust all the points of the z plane at which the equation $f(z, w) = 0$ has repeated roots or the polynomial $A_0(z)$ has a zero. In other words, the zeros of $R_w(z)$ are those and only those points of the z plane at which the number of different roots of the equation $f(z, w) = 0$ is less than n.

If a point z_0 is not a zero of the resultant, then there is a neighborhood of z_0 such that there exist n power series representing single-valued branches of an *algebraic function* $w = w(z)$, defined by the equation

$$f(z, w) = 0. \tag{I.7.3}$$

Each of these series can be continued analytically along any curve not passing through the zeros of the resultant, and each can be obtained as an analytic continuation of another along a closed curve which starts and ends at the given point z_0. Under these circumstances, as was already noticed by Puiseux, here continuation. That is, when constructing the continuation one has only to make sure that the values of the roots of equation (I.7.3) chosen one at a time as values of a *branch* of an algebraic function are arbitrarily close to each other if the points are sufficiently close. Thus an n-valued analytic function is obtained, whose singular points can only be zeros of the resultant or $z = \infty$.

If ζ_0 is a singular point and $A_0(\zeta_0) \neq 0$ at ζ_0, then by continuing in a neighborhood (which contains no other singular points) of ζ_0 we find that all n branches in this neighborhood fall into cycles, some of which may contain only a single branch. Here, if a cycle joins s branches ($s \leq n$) then all these branches can be represented in this neighborhood by an expansion of the form

$$w = \sum_{k=0}^{\infty} c_k (z - \zeta_0)^{k/s}, \qquad |z - \zeta_0| < \rho. \tag{I.7.4}$$

Thus, all the branches of this cycle have the same value c_0 at ζ_0. This value is an s-fold root of the equation $f(\zeta_0, w) = 0$ (simple if $s = 1$). If we fix the value of one of the branches of the cycle at some point z_0 of the neighborhood $|z - \zeta_0| < \rho$ and then extend the branch continuously over the circle $|z - \zeta_0| = |z_0 - \zeta_0|$, then on each circuit of the circle in the same direction we will reach the point z_0 with new values belonging to other branches of the same cycle until after s circuits we return to the original value of the original branch.

Therefore, to a given cycle there corresponds a *branch point* of order $s - 1$ at ζ_0 (if $s > 1$). For a point ζ_0 on the plane there could be as many of these points as there are distinct cycles containing the point, such that values $s > 1$ correspond to these cycles (obviously the number of such cycles cannot exceed $n/2$).

If $A_0(\zeta_0) = 0$, then we make the substitution $w = 1/w'$ and get

$$A_n w'^n + A_{n-1} w'^{n-1} + \cdots + A_1 w' + A_0 = 0. \tag{I.7.5}$$

One or several roots become zero at the point ζ_0. The corresponding branches in the neighborhood of ζ_0 fall into cycles of the form

$$w' = \sum_{k=p}^{\infty} c_k (z - \zeta_0)^{k/s}, \qquad (p \geq 1,\ c_p \neq 0).$$

We stress that we selected only cycles that join branches such that these cycles do not take the value zero at the point ζ_0. Returning to the variable w we obtain cycles of the form

$$w = \sum_{k=-p}^{-\infty} \gamma_k (z - \zeta_0)^{k/s}, \qquad (\gamma_{-p} \neq 0),$$

branches of which have the common value ∞ at the point ζ_0. Depending on whether $s = 1$ or $s > 1$ for a given cycle, we obtain either a *pole* at this point or again a *branch point of order* $s - 1$ (the function takes the value infinity at this point).

In order to investigate the behavior of an algebraic function at the point at infinity, we make the substitution $z = 1/z'$. In the neighborhood of $z' = 0$ we get the equation

$$a_0(z') w^n + \cdots + a_n(z') = 0,$$

where $a_j(z')$ are polynomials in z' ($a_0(z') \not\equiv 0$, $a_n(z') \not\equiv 0$). Depending on whether we have:

(1) $a_0(0) \neq 0$ and the equation has no repeated roots,
(2) $a_0(0) \neq 0$ and the equation has repeated roots, or finally
(3) $a_0(0) = 0$,

we will have one of the cases outlined above for $w = w(1/z')$. Any of them gives (using $z = 1/z'$ again) a corresponding pattern of behavior of the algebraic function at infinity. We leave it to the reader to write $w = w(z)$ as a series in each of the corresponding cases. We only need to notice that when $z = \infty$ the algebraic function can only have regular points, poles, or branch points of finite order.

What was said above characterizes the properties of the class of algebraic functions completely (Puiseux). Namely, *if a function admits an unbounded analytic continuation everywhere along any path in the extended plane (sphere) where the path does not pass through a finite number of points at which the function can have singularities; if these points can only be poles or branch points of finite order (where the function may take the value ∞) and if, finally, the number of distinct values of the function at any point of the plane does not exceed a certain positive integer n, then the function is algebraic and satisfies an equation of degree n in w.*

It is useful to study algebraic functions on the corresponding *Riemann surfaces.* In order to make a model of a Riemann surface we mark all the points $\zeta_1, \dots, \zeta_N$ of the z plane that correspond to singular points of the algebraic function. From each of these points we make a cut along a ray or a broken line toward $z = \infty$ so that these cuts do not intersect. We now make n identical copies of this plane and on each of them we define a single-valued branch of the function (branches are distinct for different sheets) by continuation starting from one (fixed at the same regular point z_0) of the values of the function. We can enumerate all the branches $w_1, \dots, w_n$ and give the same numbers to corresponding sheets.

The cuts on the jth sheet will in general be *lines of discontinuity* for the jth branch. Branch w_j will have the same values on the different *sides of the cut* $\mathscr{L}_k$ starting from a point ζ_k only when the branch is a part of a cycle corresponding to the point ζ_k and when this cycle contains only this branch alone. In this case we eliminate the cut $\mathscr{L}_k$ on the jth sheet, leaving only those cuts for which w_j has unequal values on the two sides of the cut. This operation is conducted for every sheet.

It remains now to glue the opposite sides of the cuts with the same name pairwise by putting the sheets over each other and obeying the following rule: a side of the cut $\mathscr{L}_k$ on a jth sheet is glued to the opposite side of a cut with the same name on a sheet j' if and only if w_j, when continued in a neighborhood of ζ_k from one side to the other, becomes $w_{j'}$ directly (and thus both branches w_j and $w_{j'}$ belong to the same cycle corresponding to ζ_k). Performing everything mentioned above we then obtain the closed n-sheeted Riemann surface F of the function $w(z)$. This algebraic function will be a single-valued function at each point of the surface thus constructed. The method of construction of F described above can be modified (see section I.8), but the result will be the same: the same number of sheets, all the same branch points as far as number, position, and order of branching is concerned, the same connection between sheets in a neighborhood of each branch point.

§8. Abelian integrals. Abel's theorem

Given an algebraic function $w(z)$ and its Riemann surface F defined by an irreducible algebraic equation

$$f\left(\overset{m}{z}, \overset{n}{w}\right) = 0, \tag{I.8.1}$$

one can develop on it a theory analogous to the theory of rational functions and their integrals. The latter theory then corresponds to the special case when $n = 1$ and F is the extended plane (sphere).

First of all, for each rational function of two variables $R(z, \zeta)$ we define an algebraic function $R[z, w(z)] = R(z, w)$. It is obviously single-valued on F. The fact that it is indeed algebraic can be checked either by noticing that as a function of the point z on the plane it is no more than n-valued and that all its singular points are algebraic (poles or branch points of finite order), or by eliminating w from the equations (I.8.1) and (I.8.2)

$$W = R(z, w). \tag{I.8.2}$$

For any rectifiable (in the spherical metric) curve Λ on F an integral

$$I = \int_\Lambda R(z, w)\, dz. \tag{I.8.3}$$

can be defined in the usual way (as the limit of sums). Such an integral is called *Abelian* in honor of N. H. Abel, who first obtained general results about integrals of the type (I.8.3) (see *Abel's theorem* below). It can be shown that if Λ is a closed curve which can be deformed continuously into a point on the surface F in such a way that under this deformation it does not pass through points at which both the integral and $R(z, w)$ take the value ∞, then

$$I = \int_\Lambda R(z, w)\, dz = 0. \tag{I.8.4}$$

This is the *Cauchy's Integral Theorem for a function on a Riemann surface* F.

If the deformation mentioned above is impossible without touching the points where the integral and $R(z, w)$ are ∞, then the integral is equal to the product of $2\pi i$ and the sum of the residues of the function, where the residues are computed appropriately for the points mentioned.

Finally, if Λ cannot be deformed continuously into a point on F, then the value ω, which can be zero or nonzero, of the integral along Λ remains the same under all the continuous deformations mentioned above. In the case when $\omega \neq 0$ it is called a *period of the Abelian integral.* Here the term *period* has a different meaning from that used above. More precisely, a period of an Abelian integral is a period (in the usual sense of the word) of the function inverse to this integral.

If we fix only the endpoints z_0 and z on the curve Λ (not closed in general), then the integral

$$\int_\Lambda R(z, w)\, dz = \int_{z_0}^{z} R(z, w)\, dz \tag{I.8.3'}$$

will be a multiple-valued function not only of z but also of the corresponding point on the Riemann surface. Indeed, on two curves Λ_1 and Λ_2, such that one cannot be reduced to the other by the method explained above, the integral takes different values. If we denote its value along some curve Λ_0 by I_0, then any value of the integral (I.8.3′) is obtained from $I_0(z)$ by adding an integral linear combination of a finite number of independent periods $\omega_1, \dots, \omega_\nu$,

$$I(z) = m_1\omega_1 + \cdots + m_\nu\omega_\nu + I_0(z). \tag{I.8.5}$$

Thus, the nature of the multiple-valuedness of the Abelian integrals is determined by the topological properties of the surface F, since these properties determine the existence of closed curves which cannot be reduced to a point as well as the number of such curves (i.e., more precisely, the number of independent classes of such curves). This is why the main topological invariant of a closed one-sided surface F, namely its *genus* (usually denoted by p), plays such an important role in the theory of Abelian integrals. The genus can be determined from a graph as the number of "holes" or "handles" of a surface in its canonical representation (Figure 5).

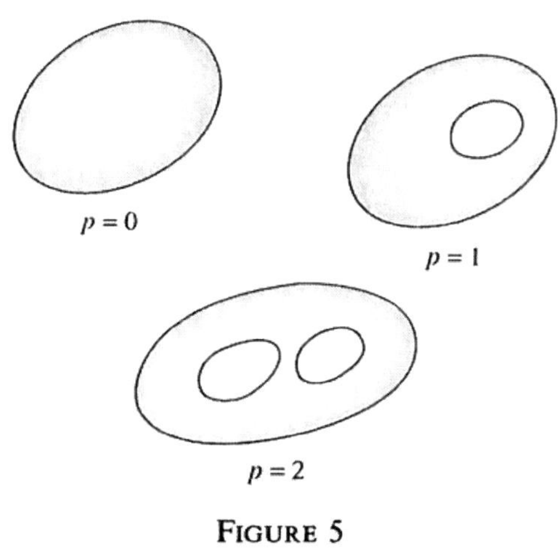

FIGURE 5

The notion of genus is also transferred to the corresponding algebraic equation $f(\overset{m}{z}, \overset{n}{w}) = 0$: one calls this equation an equation of genus p. Genus is a much deeper characteristic of an algebraic equation than the degrees m and n. Assume that z and w are transformed using the formulas $z = R_1(z', w')$ and $w = R_2(z', w')$, where R_1 and R_2 are rational functions, and let $\tilde{f}\left(\overset{m'}{z'}, \overset{n'}{w'}\right) = 0$ be the transformed equation. Of course, the degrees m and n as well as the genus p will then change. But the genus stays the same if the transformation is *birational*, i.e., if using the transformation formulas and the equation expressing the relation between z and w one can express z' and w' in the form of functions of z and w: $z' = S_1(z, w)$, $w' = S_2(z, w)$ which are also rational.

We will illustrate everything said above by using an integral considered by Jacobi (see section I.5) as a particular example. First we present the necessary definitions. An Abelian integral is called an *integral of the first kind* if its value remains finite when integration is carried out along any curve on the surface. *A hyperelliptic* integral is an integral of the special type

$$I(z) = \int_{z_0}^{z} R\left(z, \sqrt{\mathfrak{X}(z)}\right) dz, \tag{I.8.6}$$

where $\mathfrak{X}(z)$ is a polynomial of degree m (not less than five) without repeated zeros. By simple transformations a hyperelliptic integral can be expressed in the form

$$I(z) = \int_{z_0}^{z} \frac{q(z)\, dz}{r(z)\sqrt{\mathfrak{X}(z)}}, \tag{I.8.7}$$

(to convince yourself that this is true it suffices to notice that any polynomial in z and $\sqrt{\mathfrak{X}(z)}$ can be transformed into the form $P_1(z) + P_2(z)\sqrt{\mathfrak{X}(z)}$, where P_1 and P_2 are polynomials).

If $q(z)$ and $r(z)$ are relatively prime, then $I(z)$ is an integral of the first kind if and only if, first, $r(z)$ has no roots, i.e., $r(z) = \text{const}$ (we set $r(z) = 1$); and, second, $m/2$ exceeds the degree of $q(z)$ by more than 1. The last condition is necessary to guarantee the existence of the improper integral

$$\int_{z_0}^{\infty} \frac{q(z)\, dz}{\sqrt{\mathfrak{X}(z)}}.$$

Independently of whether the degree of $\mathfrak{X}(z)$ is even ($m = 2\mu$) or odd ($m = 2\mu - 1$) there are $p = \mu - 1 = [(m-1)/2]$ linearly independent integrals

$$\int \frac{dz}{\sqrt{\mathfrak{X}}}, \int \frac{z\, dz}{\sqrt{\mathfrak{X}}}, \dots, \int \frac{z^{p-1}}{\sqrt{\mathfrak{X}}}\, dz \tag{I.8.8}$$

that satisfy the condition given above. Every integral of the first kind can be expressed as a linear combination of these integrals.

In the case of hyperelliptic integrals, the algebraic function $w(z)$ is defined by an equation of an especially simple form

$$f\left(\overset{m}{z}, \overset{2}{w}\right) = w^2 - \mathfrak{X}(z) = 0. \tag{I.8.9}$$

Here all the branch points are among the zeros of the polynomial $\mathfrak{X}(z)$, $\zeta_1, \zeta_2, \ldots, \zeta_m$, but in the case when m is odd, one more branch point exists at infinity. The order of each branch point is 1 and the total number is $2p + 2$ $(p = [(m-1)/2])$.

For example, in the case of the hyperelliptic integral of Jacobi (see section I.5) where

$$\mathfrak{X}(z) = z(1-z)(1-\varkappa^2 z)(1-\lambda^2 z)(1-\mu^2 z), \qquad 1 > \varkappa^2 > \lambda^2 > \mu^2 > 0,$$

all the branch points are on the real axis and are $0, 1, \frac{1}{\varkappa^2}, \frac{1}{\lambda^2}, \frac{1}{\mu^2}$ and ∞. Here $p = [(5-1)/2] = 2$, and there are $2p+2 = 6$ branch points and two linearly independent integrals of the first kind

$$\int_0^z \frac{dz}{\sqrt{\mathfrak{X}}} \quad \text{and} \quad \int_0^z \frac{z\,dz}{\sqrt{\mathfrak{X}}}.$$

The integral considered by Jacobi, namely

$$\int_0^z \frac{(\alpha + \beta z)\,dz}{\sqrt{\mathfrak{X}}}$$

is the most general hyperelliptic integral of the first kind for the case of a polynomial $\mathfrak{X}$ of 5th, as well as 6th, degree.

In order to represent the facts about Abelian integrals graphically we start with the construction of the Riemann surface of the function $w = \sqrt{\mathfrak{X}(z)}$ defined by the equation

$$f\left(\overset{5}{z}, \overset{2}{w}\right) = w^2 - z(1-z)(1-\varkappa^2 z)(1-\lambda^2 z)(1-\mu^2 z) = 0. \tag{I.8.10}$$

We will verify in particular that the genus of F is equal to the number $p = [(m-1)/2] = 2$ of linear independent integrals of the first kind. A similar statement is also true for any algebraic function $w(z)$, but the previous formula relating p and m does not hold in the general case.

We choose a method of construction of F which is somewhat different from the method presented in section I.7. Instead of two copies of the plane we take two copies of the sphere . We mark on each of them the branch points $\zeta_1 = \infty, \zeta_2 = 0, \zeta_3 = 1, \zeta_4 = 1/\varkappa^2, \zeta_5 = 1/\lambda^2$, and $\zeta_6 = 1/\mu^2$ and connect them pairwise (ζ_1 with ζ_2, ζ_3 with ζ_4, and ζ_5 with ζ_6) by cuts. One of the two identical copies of the sphere is shown in Figure 6. We also construct six closed rectifiable curves Λ_j $(j = 1, 2, 3, 4, 5, 6)$ such that any Λ_j can be continuously deformed into a segment $[\zeta_j, \zeta_{j+1}]$ $(\zeta_7 = \zeta_1)$ without passing through the points ζ_j (see Figure 6 again). Here the curves $\Lambda_1, \Lambda_3, \Lambda_5$ encircling the cuts can be considered to lie on the same copy

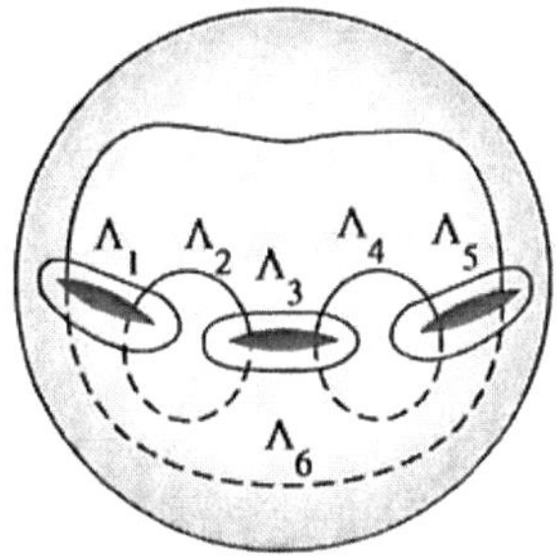

FIGURE 6

of the sphere (they are shown as solid lines on the figure) and the curves Λ_2, Λ_4, Λ_6 which pass from one side of a cut to the other lie partially on the same copy of the sphere (solid line) and partially on the other copy of the sphere (dotted line). The integrals along Λ_j,

$$\int_{\Lambda_j} \frac{\alpha + \beta z}{\sqrt{\mathfrak{X}(z)}}\, dz = \omega_j \qquad (j = 1, 2, 3, 4, 5, 6),$$

are then the periods of the hyperelliptic integral under consideration. They coincide (up to sign) with the periods found by Jacobi (I.5.5).

We now connect the two copies of the sphere to form a two-sheeted Riemann surface F by gluing the corresponding sides of the cuts to each other. However, to represent F directly in canonical form, we will not superimpose one sphere on the other as two layers but put one against the other (cut to cut) and identify the edges of the cuts which need to be glued.

Here (for better visual clarity) we enlarge the cuts to ovals and stretch the adjacent parts of spheres toward each other so that they have a tubular form. After gluing, we finally obtain the desired surface F (Figure 7).

It is clear, first, that its genus is 2, i.e., that it is the same as $p = [(m-1)/2]$. Furthermore, the number of distinct periods of the integral $I(z)$ which are independent of each other is not 6, but $4 = 2p$. This is not an accident, but a manifestation of a general regularity. This fact, as we saw in the section I.5, was proved by Jacobi himself. But here we will be able to display the two

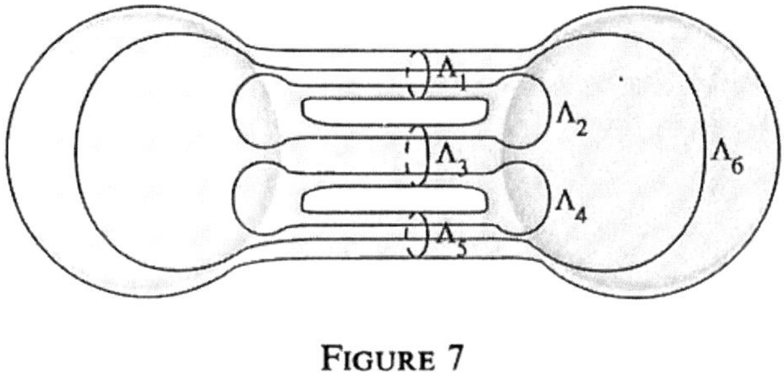

FIGURE 7

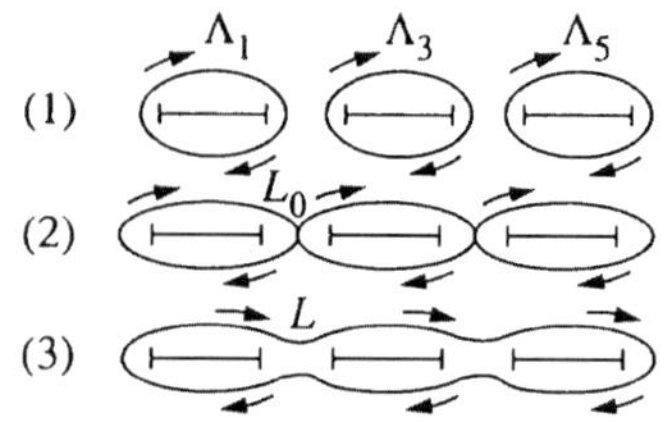

FIGURE 8

relations that he found between the periods of the equality (I.5.6) by using a different method, which in the end is based on Cauchy's integral theorem. Using the relation between the curves Λ_2, Λ_4, and Λ_6 (Figure 6 or Figure 7), we see that

$$\int_{\Lambda_2} + \int_{\Lambda_4} + \int_{\Lambda_6} = 0, \qquad \text{i.e., } \omega_2 + \omega_4 + \omega_6 = 0;$$

this is the first linear relation between the periods. We can further imagine that the curves Λ_I, Λ_3, and Λ_5 grow until they coincide (Figure 8 gives several stages of this deformation). Then $\int_{\Lambda_1} + \int_{\Lambda_3} + \int_{\Lambda_5} = \int_{L_0}$ holds and on the other hand $\int_{L_0} = \int_L = 0$ is true, i.e., $\omega_I + \omega_3 + \omega_5 = 0$ holds, and this is the second linear relation between the periods.

Thus only 4 periods can be considered independent, for example ω_I, ω_2, ω_5, and ω_6, and just this choice was made by Jacobi.

Because of the formula (I.8.5) which gives the set of all values of the Abelian integral $I(z)$ we have in this case

$$\int_0^{z_0} \frac{\alpha + \beta(z)}{\sqrt{\mathfrak{X}(z)}}\, dz = I_0 + m\omega_I + m'\omega_2 + m''\omega_5 + m'''\omega_6,$$

where m, m', m'', and m''' are any integers.

In the summer of 1826 Abel obtained a quite general statement (which now bears his name) about Abelian integrals (it was published in complete form only in 1841, 12 years after his death). We will not use it in further developments but we will state it in its entirety, and give the proof (as was done by Abel himself) only for the case of Abelian integrals of the first kind, thus simplifying the theorem. We will see that in this case Abel's theorem can be considered as a sweeping generalization of Euler's result (considered in §I.1) about elliptic integrals.

We will treat the equation

$$f\left(\overset{m}{z}, \overset{n}{w}\right) = 0, \tag{I.8.11}$$

defining the algebraic function $w = w(z)$ as the equation of an algebraic curve, and also consider the algebraic curve

$$\varphi\left(\overset{\mu}{z}, \overset{\nu}{w}\right) = 0. \tag{I.8.12}$$

Let the coefficients of the polynomial $\varphi(z, w)$ (all or only some) be variable and denote the coefficients by $a_1, \dots, a_k$. Then equation (I.8.12) gives a family of curves, and the points of intersection (I.8.11) and (I.8.12) are functions of the parameters a_j $(j = 1, \dots, k)$. Denote these points by (z_i, w_i) $(i = 1, \dots, N)$. Then the sum of the values of the Abelian integrals of the function $R(z, w)$, taken from the same point (z_0, w_0) on the curve (I.8.11) (z_0 is not a singular point of $w(z)$) to the points of intersection indicated above, will be a function of a_j:

$$\sum_{i=1}^{N} \int_{z_0}^{z_i} R(z, w)\, dz = \psi(a_1, \dots, a_k). \tag{I.8.13}$$

Notice that by fixing the values of w_0 at the point z_0 we choose one of the n single-valued branches of $w(z)$ in the neighborhood of z_0 and continue it along the integration paths. Actually, this function is in principle multiple-valued because Abelian integrals are multiple-valued, and the function is defined up to an arbitrary period (integral linear combination of basic periods). However, we fix the paths along which the integrals are taken for certain initial values of the parameters $a_j = a_j^{(0)}$ and later change the paths continuously by gradually extending them (as the parameters change continuously) by arcs extending to the endpoints z_i.

We also assume, for simplicity, that the paths neither pass through the zeros of the resultant (I.7.2) of the equations $f = 0$ and $\partial f/\partial w = 0$ nor through the singular points of the integrand $R(z, w)$.

Abel's theorem is the assertion that *the function* (I.8.13) *is in all cases a sum of rational functions and logarithms of rational functions, multiplied by constants.* It follows, in particular, that for some values of the parameters (possibly infinite) these functions will become infinite unless ψ is a constant. However, in the case when an Abelian integral $\int R(z, w)\, dz$ is an integral of the first kind, an infinite value for the sum of integrals is not possible because of the very definition of an integral of the first kind. It follows that *for integrals of the first kind the sum* (I.8.13) *is a constant.* Here we will prove this very special case of Abel's theorem (actually, the proof of the general case differs little from the proof given below).

We denote by δ the differentials of the functions, subject to variations of the parameters a_j. We have

$$\delta\psi = \sum_{i=1}^{N} R(z_i, w_i)\delta_{z_i}. \tag{I.8.14}$$

If we substitute z_i and w_i into the equations (I.8.11) and (I.8.12) and differentiate the result, we obtain

$$\frac{\partial f}{\partial z_i}\delta z_i + \frac{\partial f}{\partial w_i}\delta w_i = 0, \qquad \frac{\partial \varphi}{\partial z_i}\delta z_i + \frac{\partial \varphi}{\partial w_i}\delta w_i + \delta\varphi_i = 0. \tag{I.8.15}$$

Because of our assumptions we have $\partial f/\partial w_i \neq 0$ (otherwise we would have $f = 0$, $\partial f/\partial w = 0$ and therefore $R_w = 0$). Hence $\delta w_i = -(\frac{\partial f}{\partial z_i}/\frac{\partial f}{\partial w_i})\delta z_i$ and from the second of the equations (I.8.15) we get

$$\delta_{z_i} = \frac{\partial f/\partial w_i}{(\partial f/\partial z_i)(\partial \varphi/\partial w_i) - (\partial f/\partial z_i)(\partial f/\partial w_i)}\delta\varphi_i \tag{I.8.16}$$

(provided we assume that the Jacobian $D(f, \varphi)/D(z, w) \neq 0$ at the intersections of the curves (I.8.11) and (I.8.12), i.e., that the curve $\varphi(z, w) = 0$ does not have a tangent in common with $f(z, w) = 0$ at any of these points). Substituting (I.8.16) into (I.8.14), we obtain

$$\delta\psi = \sum_{i=1}^{N} \frac{\partial f R(z_i, w_i)/\partial w_i}{(\partial f/\partial z_i)(\partial\varphi/\partial w_i) - (\partial\varphi/\partial z_i)(\partial f/\partial w_i)}\delta\varphi_i. \tag{I.8.17}$$

Since $\delta\varphi_i = \sum_{j=1}^{k}(\partial\varphi(z_i, w_i)/\partial a_j)\delta a_j$ it can easily be discovered that the coefficients of δa_j in the expression for $\delta\psi$, i.e., the derivatives $\partial\psi/\partial a_j$, are *symmetric functions of the roots* z_i, w_i $(i = 1, \dots, N)$ of the system of equations (I.8.11), (I.8.12). Therefore they are rational functions of the coefficients of the equations and, in particular, of the coefficients $a_1, \dots, a_k$. Thus

$$\delta\psi = \sum_{j=1}^{k} r_j(a_1, \dots, a_k)\delta a_j. \tag{I.8.18}$$

This, then, is the main part of the proof of Abel's theorem in the form given by Abel himself. It remains only to note that for the integrals of the first kind the rational functions r_j must be finite everywhere, i.e., must be constants. But then ψ is a linear function of the parameters and remains bounded even under the condition that some or all of the parameters tend to infinity. From this it finally follows that $\psi \equiv \text{const}$. Thus for an integral of the first kind $\int R(z, w)\,dz$ we have

$$\sum_{i=1}^{N}\int_{z_0}^{z_i} R(z, w)\,dz \equiv \text{const}, \tag{I.8.19}$$

where (z_i, w_i) $(i = 1, \dots, N)$ are the points of intersection of the curve (I.8.11) with the family of curves (I.8.12).

We will show (following Briot and Bouquet) that Euler's addition theorem (section I.1) for elliptic integrals can be obtained as a very special case of Abel's theorem. The algebraic function $w = \sqrt{1 + mz^2 + nz^4}$ is defined by the equation

$$f\left(\overset{4}{z}, \overset{2}{w}\right) = w^2 - (1 + mz^2 + nz^4) = 0. \tag{I.8.20}$$

Consider the points of intersection of the curve (I.8.20) with the family of "parabolas"

$$\varphi\left(\overset{2}{z}, \overset{1}{w}\right) = w - (1 + pz + qz^2) = 0 \tag{I.8.21}$$

passing through the point $z = 0$, $w = 1$ of this curve (p and q are arbitrary parameters). Here the corresponding values of z_j $(j = 1, 2, 3, 4)$ are the roots of the equation

$$(q^2 - n)z^4 + 2pqz^3 + (p^2 + 2q - m)z^2 + 2pz = 0,$$

or, if we eliminate the obvious root $z_4 = 0$, of the equation

$$(q^2 - n)z^3 + 2pqz^2 + (p^2 + 2q - m)z + 2p = 0. \tag{I.8.22}$$

From this we obtain

$$z_1 + z_2 + z_3 = -\frac{2pq}{q^2 - n}, \qquad z_1 z_2 z_3 = -\frac{2p}{q^2 - n},$$

and therefore $(z_1 + z_2 + z_3) - qz_1z_2z_3 = 0$, i.e.,

$$z_3 = \frac{z_1 + z_2}{qz_1z_2 - 1}. \tag{I.8.23}$$

We transform this expression by using the equations (I.8.21) and (I.8.20). First we derive $z_1w_2 - z_2w_1 = -(qz_1z_2 - 1)(z_1 - z_2)$ from (I.8.21) and therefore

$$z_3 = -\frac{z_1^2 - z_2^2}{z_1w_2 - z_2w_1}. \tag{I.8.24}$$

Furthermore, equation (I.8.20) gives us $z_1^2w_2^2 - z_2^2w_1^2 = (z_1^2 - z_2^2)(1 - nz_1^2z_2^2)$. If we determine $z_1^2 - z_2^2$ from this, and substitute the result into (I.8.24), we find

$$z_3 = -\frac{z_1w_2 + z_2w_1}{1 - nz_1^2z_2^2}. \tag{I.8.25}$$

Because of the general relation (I.8.19) we have the following equation when z_0 is taken to be zero:

$$\int_0^{z_1} \frac{dz}{\sqrt{1 + mz^2 + nz^4}} + \int_0^{z_2} \frac{dz}{\sqrt{1 + mz^2 + nz^4}} + \int_0^{z_3} \frac{dz}{\sqrt{1 + mz^2 + nz^4}} = \text{const.} \tag{I.8.26}$$

To determine this constant we notice that in this case all the points z_1, z_2, z_3 simultaneously coincide with the origin for certain values of the parameters. It follows that the right-hand side of (I.8.26) is equal to zero. Indeed, set $p = 0$ and $q = m/2$. Then equation (I.8.22) takes the form $(\frac{m^2}{4} - n)z^3 = 0$, from which it follows that $z_1 = z_2 = z_3 = 0$.

In order to obtain Euler's result, expressed by (I.1.12), we substitute $-z$

for z in the integral $\int_0^{z_3}$. We find

$$\int_0^{z_1} \frac{dz}{\sqrt{1+mz^2+nz^4}} + \int_0^{z_2} \frac{dz}{\sqrt{1+mz^2+nz^4}} = \int_0^{-z_3} \frac{dz}{\sqrt{1+mz^2+nz^4}} = \int_0^{(z_1w_2+z_2w_1)/(1-nz_1^2z_2^2)} \frac{dz}{\sqrt{1+mz^2+nz^4}},$$

and this is (I.1.12) in a somewhat different notation.

§9. Main directions of development of the theory of Abelian functions

We can now state Jacobi's inversion problem in the most general form. Let

$$f\left(\overset{m}{z}, \overset{n}{w}\right) = 0 \tag{I.9.1}$$

be an irreducible algebraic equation of genus p and let

$$\int_{a_1}^{z_1} R_1(z, w)\,dz, \dots, \int_{a_p}^{z_p} R_p(z, w)\,dz \tag{I.9.2}$$

be the p linearly independent Abelian integrals of the first kind. We construct the p sums of their values:

$$\begin{aligned} \int_{a_1}^{z_1} R_1\,dz + \cdots + \int_{a_p}^{z_p} R_1\,dz &= u_1, \\ \cdots\cdots\cdots\cdots\cdots&\cdots\cdots \\ \int_{a_1}^{z_1} R_p\,dz + \cdots + \int_{a_p}^{z_p} R_p\,dz &= u_p. \end{aligned} \tag{I.9.3}$$

Here the right-hand sides of the equalities are multiple-valued functions of the limits $z_1, \dots, z_p$ of the integrals. They, like the integrals themselves, are determined up to the periods of the integrals. The problem is then to *study the properties of z_j as functions of the p variables $u_1, \dots, u_p$ and to construct these functions.*

We can conclude from Abel's theorem that, if not $z_j = z_j(u_1, \dots, u_p)$ themselves, at least their symmetric functions will be single-valued functions of the p variables $u_1, \dots, u_p$, and moreover they will be representable in a neighborhood of any point $u_1^{(0)}, u_2^{(0)}, \dots, u_p^{(0)}$ as the ratio of two functions which are analytic at that point. See for example.([8]) And this means that they are meromorphic functions of p complex variables.

Let $\omega_j^{(k)}$, $k = 1, \dots, 2p$ be a system of $2p$ independent periods of the

([8]) E. Picard, *Traité d'analyse*, vol. II, 3rd ed., Gauthier-Villars, 1925, pp. 510–512.

integral $\int R_j(z, w)\,dz$ $(j = 1, \ldots, p)$. We consider

$$\omega^{(k)} = \begin{pmatrix} \omega_1^{(k)} \\ \vdots \\ \omega_p^{(k)} \end{pmatrix}$$

as a p-vector in the space $\mathbf{C}^p$. If we also set $(u_1 \cdots u_p)^{\mathrm{T}}$ then it can be deduced that the values $z_1, \ldots, z_p$ do not change if $u + \omega^{(k)}$ is substituted for u in (I.9.3). In other words, the vectors $\omega^{(1)}, \ldots, \omega^{(2p)}$ form a system of $2p$ independent periods of the functions $z_1, \ldots, z_p$ and therefore of their symmetric functions (see section II.3 and the following discussion for facts about meromorphic functions of several variables and their periods).

Meromorphic functions of p complex variables which have $2p$ independent periods are called Abelian functions.

In the simplest case, $p = 1$, we have the class of elliptic functions. As we mentioned earlier, they appeared as a result of the inversion of the elliptic integral of the first kind $\int_0^z dz/\sqrt{\mathfrak{X}(z)} = u$, where $\mathfrak{X}(z)$ is a polynomial of the fourth (or third) degree. It is appropriate to recall that the historical development of the theory of elliptic functions, which had started from the point of origin that we have described, continued on as if it had forgotten about its origin very soon (after Liouville and especially after Weierstrass). The theory of elliptic functions came to be considered as *the theory of doubly periodic meromorphic functions.* The fact that they help to solve the problem of inverting elliptic integrals was then presented only as one of many applications of the general theory.

The theory of Abelian functions did not undergo a similar evolution. More precisely, in the second half of the nineteenth century it came to be studied in several different directions, each of which is still of interest and retains its scientific significance.

One of these directions (Riemann (1857) and Weierstrass (1876)) is the deep investigation of algebraic functions and Abelian integrals by means of the tools of analysis. From this point of view, Abelian functions appear (as was mentioned earlier) as functions that solve the inversion problem of Jacobi. Their properties follow from our preparatory study of Abelian integrals. Riemann, moreover, investigates algebraic functions and Abelian integrals over the Riemann surfaces which he introduced primarily for this purpose. He established the existence of Abelian integrals with certain properties as a result of the application of a general extremum principle in the theory of functions that he developed. This is called the Dirichlet principle. However, when he needed to solve the Jacobi inversion problem, he used techniques developed in the works of Göpel and Rosenhain, using theta functions of several variables (his predecessors had restricted themselves to the case of two variables). It is well known that the Dirichlet principle was sharply criticized by Weierstrass, and as a consequence it lost its credibility among mathematicians. An attempt to follow the lines of Riemann's thought without using the Dirichlet principle was made by K. Neumann in his "Lectures on Riemann's theory of Abelian integrals." ([9]) Incidentally, the credit for the real resurrection of Riemann's plan, after the rehabilitation of the Dirichlet principle by D. Hilbert, belongs to H. Weyl. ([10])

Weierstrass bases his construction of the theory of Abelian functions on the system of equations (I.9.3) itself. He treats this as the system of differential equations

$$\begin{aligned} R_1(z_1, w_1)\, dz_1 + \cdots + R_1(z_p, w_p)\, dz_p &= du_1, \\ \cdots\cdots\cdots\cdots\cdots\cdots\cdots\cdots&\cdots\cdots \\ R_p(z_1, w_1)\, dz_1 + \cdots + R_p(z_p, w_p)\, dz_p &= du_p, \end{aligned} \tag{I.9.4}$$

where z and w are related by the equation $f(z, w) = 0$. He then completely ignores Riemann surfaces. His apparatus is obtained by combining analytic (expansion into power series) and algebraic methods. He shows that the system (I.9.4) can be integrated in such a way that the symmetric functions of z_j $(j = 1, \ldots, p)$ become Abelian functions of u_k $(k = 1, \ldots, p)$. The theta functions appear in his treatment only afterwards, as the simplest entire functions in terms of which all Abelian functions can be expressed. Incidentally, Weierstrass pointed out fairly early (in 1869) that the functions that led to the solution of Jacobi's problem exhaust the entire class of Abelian functions, as it was defined above. Consequently the functions that appear in solving Jacobi's problem are called, as Poincaré suggested, *the special Abelian functions.*

It is curious to note that the only two Russian expositions of the theory of Abelian functions up to recent times were written in the spirit of Weierstrass. Both of these appeared at the end of the last century ([11]) and are mainly of historical interest.

One direction of the development of the theory of Abelian functions originates from the work of Clebsch and Gordan. ([12]) This direction is important for its ideas and methods; here there appeared the tendency to bring this theory nearer to algebraic geometry. The main theorems were considered as assertions about algebraic curves, and the methods of algebraic geometry (in particular, projective coordinates) were used to prove the theorems.

In a classical French treatise on analysis, such as the "Traité d'analyse" of Picard (see the footnote on the page 40) or in an entirely modernized but fairly traditional one, as far as concerns the part that interests us, ([13]) the

([9]) C. G. Neumann, *Vorlesungen über Riemann's Theorie der Abel'schen Integrale*, 1st ed., 1865, 2nd ed. 1884, Teubner, Leipzig.

([10]) H. Weyl, *Die Idee der Riemannschen Fläche*, 1st ed., Teubner, Leipzig 1913, 3rd ed., Teubner, Stuttgart, 1955.

([11]) M. A. Tikhomandritskiĭ, *Foundations of the theory of Abelian integrals*, Kharkov, 1895; V. P. Ermakov, *The theory of Abelian functions without Riemann surfaces*, Kiev, 1897.

([12]) A. Clebsch and P. Gordon, *Theorie der abelschen Functionen*, Teubner, Leipzig, 1866.

([13]) J. Favard, *Cours d'analyse de l'École Polytechnique*, t. II, Gauthier-Villars, Paris, 1960.

reader is presented with a synthesis of Riemann's ideas (without the Dirichlet principle) and Weierstrass's, on one hand, and the main facts and ideas of algebraic geometry on the other. A decisive step towards algebraization of the entire theory was made in a joint article by Dedekind and Weber. ([14]) As N. Bourbaki says at the end of his "Algèbre commutative" (Actualités Sci. Ind., no. 1314, Hermann, Paris, 1965, p. 124) "L'idée essentielle de leur travail est de calquer la théorie des fonctions algébriques d'une variable sur la théorie des nombres algébriques telle que venait de la développer Dedekind". We note that the new approach has led to a significant difference from a function-theoretic treatment: "autrement dit de considerer, dans l'expression $f(x)$, *f comme variable et* x *comme fixe*" (ibid, page 125, italics by N. Bourbaki).

We cannot give any further details here. As a result of the development of this direction, the theory of algebraic and Abelian functions came to be considered, to a significant extent, as an area of applications of notions and methods which today constitute a large part of commutative algebra. We mention two works where this direction is presented in modern form: C. Chevalley's "Introduction to the theory of algebraic functions of one variable." ([15]) and S. Lang's "Introduction to algebraic and abelian functions" ([2]). A distinctive feature of these books, notwithstanding their indisputable scientific merit, is that their treatments, nurtured in the lap of classical mathematical analysis, are far removed from their native environment. The advocates of algebra and algebraic geometry have done much to disguise the subject's "family" resemblance.

In our exposition we retain the classical spirit of the theory of analytic functions, but, of course, use the facts that have been obtained by the development of the basic theory of functions of several complex variables. This direction can be described as a further development of Jacobi's inversion method, with the significant difference that we will not be interested in completing the solution of the inversion problem, since our final goal is to construct the foundations of the theory of general (not special) Abelian functions. Consequently, for example, we will not base our discussion on the theory of algebraic functions and Abelian integrals. Instead, we will need some facts from the theory of analytic functions of several complex variables. To be sure, the number of these facts is quite modest (see Chapter II). An important role will be played by the Poincaré-Cousin theorem on the existence of a representation of any meromorphic function as the quotient of entire functions which are relatively prime at each point (see Appendix B). The case $p = 2$ was worked out by P. Appell "About periodic functions of two

---

([14]) R. Dedekind and H. Weber, *Theorie der algebraischen Functionen einer Veränderlichen*, J. Reine Angew. Math. **92** (1882), 181–290.

([15]) C. Chevalley, *Introduction to the theory of algebraic functions of one variable*, Amer. Math. Soc., Providence, R.I., 1951.

variables" ([16]) who used earlier ideas and results of G. Frobenius's "Foundations of the theory of Jacobian functions."([17]) Our discussion will be directly based on the excellent exposition in a posthumous edition of the lectures by the Italian mathematician Fabio Conforto, edited by V. Grebner, A. Andreotti, and M. Rosati as "Algebraic functions and algebraic geometry." ([18])

We shall cite this book by the name of its author (Conforto) and the relevant pages.

Here we will mainly use the material of the first half of the book : "Intermediate functions and existence theorems for Abelian functions." We note that a theory of Abelian functions which is similar in spirit is constructed in C. Siegel's book "Automorphic functions of several complex variables."

([16]) P. Appell, *Sur les fonctions périodiques de deux variables*, J. Math. Pures Appl. (4) 7 (1891), 157–219.

([17]) G. Frobenius, *Ueber die Grundlagen der Theorie der Jacobischen Functionen* J. Reine Angew. Math. **97** (1884), 16–48.

([18]) Fabio Conforto, *Abelsche Funktionen und algebraische Geometrie*, Springer-Verlag, 1956.

CHAPTER II

Periodic Functions of Several Complex Variables

§1. Divisibility relation for functions analytic at a point

It will be convenient to treat points of a p-dimensional complex space $\mathbf{C}^p$ as p-dimensional complex vectors. A vector $u \in \mathbf{C}^p$ will be denoted by a column matrix $u = (u_1 \cdots u_p)^{\mathrm{T}}$ where $u_1, \dots, u_p$ are complex numbers. A complex-valued function $f(u)$ is said to be *analytic at a point* $a = (a_1 \cdots a_p)^{\mathrm{T}}$ if there exists a power series

$$f(u) = \sum c_{j_1 \cdots j_p}(u_1 - a_1)^{j_1} \cdots (u_p - a_p)^{j_p} = \sum_{(j)} c_j (u - a)^j, \qquad \text{(II.1.1)}$$

that converges to $f(u)$(uniformly and absolutely) in some p-disk $|u_k - a_k| < \rho_k$, $\rho_k > 0$, $k = 1, \dots, p$. This condition is equivalent to the requirement that the function $f(u)$ is defined and differentiable in a neighborhood of the point a, i.e., at each point u of this neighborhood there exists a total differential

$$df = \sum_{k=1}^{p} \frac{\partial f}{\partial u_k} du_k$$

(which is equal by definition to the dominant part of the corresponding increment Δf). Collecting the terms of (II.1.1) with equal sums of exponents into homogeneous polynomials, we represent the expansion (II.1.1) in the form

$$f(u) = \sum_{n=0}^{\infty} p_n(u - a),$$

where $p_0 = c_{0\cdots 0}$, $p_1(u - a) = c_{10\cdots 0}(u_1 - a_1) + \cdots + c_{0\cdots 01}(u_p - a_p)$ and so on. If $f(a) \neq 0$ then $\varphi(u) = 1/f(u)$ is also analytic at the point a. A point a at which $f(a) = 0$ but $f(u) \not\equiv 0$ is called a *zero* of $f(u)$. In this case $p_0 = 0$ but among the polynomials $p_n(u - a)$ $(n \geq 1)$ there is a polynomial of the least degree $\nu \geq 1$ not identical to zero. The number ν is called the *lower order* of the function $f(u)$ at the point a. We subject u to a linear transformation with matrix $A = (a_{jk})$, $j = 1, \dots, p$, $k = 1, \dots, p$, such that the determinant $|A| \neq 0$. Then u is transformed into $u^* = Au$ and the function $f(u)$ into $f^*(u^*) = f(u)$. It is easy to check that $f^*(u^*)$ is analytic at the point $a^* = Aa$ and moreover its lower order is still equal to ν.

If $f_1(u)$ and $f_2(u)$ are also analytic at a and if

$$f(u) = f_1(u) f_2(u), \tag{II.1.2}$$

they are called *divisors* of $f(u)$ (at this point) and the formula (II.1.2) is called a factorization of $f(u)$; for brevity we write $f_1|f$ and $f_2|f$. It is obvious that the sum of the lower orders of the divisors is equal to ν (here the lower order of a function for which a is not a zero is defined to be zero). The divisors different from zero at the point a are called **units**. Any unit is a divisor of each function $f(u)$ that is analytic at a.

Let for simplicity $a = 0$ and let $p_\nu(u)$ have the form

$$p_\nu(u) = \pi_0 u_1^\nu + \pi_1 u_1^{\nu-1} + \cdots + \pi_\nu \qquad (\nu \geq 1),$$

where π_j are homogeneous polynomials of degree j with respect to $u_2, \ldots, u_p$ and $\pi_0 \neq 0$. The *Weierstrass preparation theorem* (an essential part of it was already known to Cauchy) states that *there is then a factorization for $f(u)$ of the form*

$$f(u) = \left(\Pi_0 u_1^\nu + \Pi_1 u_1^{\nu-1} + \cdots + \Pi_\nu\right) f_1(u), \tag{II.1.3}$$

where $\Pi_0 = \pi_0$ (a constant not equal to zero), Π_j are functions of the variables $u_2, \ldots, u_p$, analytic at the origin, whose lower orders are no less than their indices, and finally where $f_1(u)$ is a unit. ([19]) An expression of the form $\sum \Pi_k u_1^{\nu-k}$ is called *a pseudopolynomial.*

We notice that if a polynomial $p_\nu(u)$ does not satisfy the preceding condition, i.e., if no power u_k^ν is present in the decomposition $p_\nu(u)$ with a nonzero coefficient, then there always is a linear transformation of the variables $u^* = Au$ transforming $f(u)$ into the required form. In fact, let

$$p_\nu(u) = \sum_{j_1+\cdots+j_p=\nu} c_{j_1\cdots j_p} u_1^{j_1} \cdots u_p^{j_p} \not\equiv 0.$$

Set $u_k = \lambda_k(u_1^* + \cdots + u_k^*)$, $\lambda_k \neq 0$, $k = 1, \ldots, p$. After the substitution we get a homogeneous polynomial $p_\nu^*(u^*)$ of degree ν in which the coefficient of $(u_1^*)^\nu$ is equal to $\sum c_{j_1\cdots j_p} \lambda_1^{j_1} \cdots \lambda_p^{j_p} = p_\nu(\lambda)$. It will be different from zero if $p_\nu(\lambda_1, \ldots, \lambda_p) \neq 0$. Because one can choose numbers $\lambda_1, \ldots, \lambda_p$ which satisfy the last condition in an infinite number of ways it follows that for any finite set of functions $f_1(u), \ldots, f_m(u)$ which have a zero at the origin, one can find a linear transformation $u^* = Au$ such that, in terms of the new variables, each of the functions will be expressed in the form (II.1.3). This will be assumed in what follows.

([19]) See, for example, A. I. Markushevich, *Theory of analytic functions*, vol. I, "Nauka", Moscow, 1967, p. 472ff. (for $p = 2$) or B. V. Shabat, *Introduction to complex analysis*, part II, "Nauka", Moscow, 1976, pp. 113–115.

Let $f(u) = f_1(u) \cdots f_m(u)$ where $f_j(u)$ $(j = 1, \dots, m)$ are not units at the point $u = 0$ (i.e., $f_j(0) = 0$). Then by what we have said,

$$f_j(u) = \left(\Pi_0^{(j)} u_1^{\nu_j} + \cdots + \Pi_{\nu_j}^{(j)}\right) \varphi_j(u), \tag{II.1.4}$$

where $\varphi_j(u)$ is a unit and therefore

$$\Pi_0 u_1^\nu + \cdots + \Pi_\nu = \left(\Pi_0^{(1)} u_1^{\nu_1} + \cdots + \Pi_{\nu_1}^{(1)}\right) \cdots \left(\Pi_0^{(m)} u_1^{\nu_m} + \cdots + \Pi_{\nu_m}^{(m)}\right) \psi(u),$$

where $\psi(u)$ is also a unit. Thus *the pseudopolynomials* $\Pi_0^{(j)} u_1^{\nu_j} + \cdots + \Pi_{\nu_j}^{(j)}$, *which characterize the divisors of the function* $f(u)$, *are themselves divisors of the pseudopolynomial in the decomposition* (II.1.3) *of this function.* This remark allows one to build a theory of divisibility for the functions analytic at a point, analogous to the theory of divisibility of polynomials of one variable (based on the Euclidian algorithm). In particular, the following theorem is true.

If $f(u)|g(u)h(u)$ *and* $f(u)$ *and* $g(u)$ *are relatively prime* (i.e., have no common divisors except units), *then* $f(u)|h(u)$.

We will not distinguish between divisors that differ by a unit factor.

A function $g(u)$ is called *prime* (at a point $u = 0$) if it is not a unit and does not have any divisors except itself and units. It is possible to prove the following assertion about the uniqueness of the factorization.

Each function $f(u)$ *factors into primes in a unique way.* In other words, if

$$f(u) = g_1(u) \cdots g_m(u) = h_1(u) \cdots h_n(u),$$

and if moreover the functions $g_i(u)$, $h_j(u)$ are prime at the point $u = 0$, then $n = m$ and the numbering of the functions $h_j(u)$ can be changed so that the conditions

$$g_j(u) = h_j(u)\varphi_j(u), \qquad j = 1, \dots, m,$$

(where $\varphi_j(u)$ are units) hold. We note that m is not larger than the lower order ν of the function $f(u)$. This follows from the fact that the lower orders of the functions $g_j(u)$ are added to each other when the functions are multiplied.

The following important conclusion follows from the Weierstrass preparation theorem: *when* $p > 1$, *the zeros of an analytic function* $f(u)$ *are not isolated; in a neighborhood of each zero* $(u = 0)$ *they form a manifold defined by an equation of the form*

$$\Pi_0 u_1^\nu + \Pi_1 u_1^{\nu-1} + \cdots + \Pi_\nu = 0. \tag{II.1.5}$$

This means that to each point in a sufficiently small $(p-1)$-dimensional neighborhood of the point $u_2 = 0, \dots, u_p = 0$ there correspond precisely ν zeros of $f(u)$ belonging to that neighborhood of the point $u = 0$. Therefore *the complex dimension of the manifold of zeros of an analytic function in a neighborhood of each zero is equal to* $p-1$.

In the case $p = 1$ the whole preceding theory becomes trivial. Here $p_n(u) = c_n u^n$. Functions that are prime at the point $u = 0$ are functions that have a simple zero:

$$f(u) = c_1 u + c_2 u^2 + \cdots, \qquad c_1 \neq 0.$$

These functions differ from each other by a factor which is equal to a unit and hence each of them can be reduced essentially to a single prime function $f(u) = u$. Hence if $f(u)$ has a zero of order ν $(\nu \geq 1)$ at the origin, its decomposition into prime factors has the form $f(u) = u^\nu \varphi(u)$, where $\varphi(u) = c_\nu + c_{\nu+1} u + \cdots$, and $c_\nu \neq 0$ is a unit.

For more details about the divisors of an analytic function of several variables see Appendix B.1 at the end of the book.

§2. Entire and meromorphic functions

A function $f(u)$ is called entire if there is a power series expansion

$$f(u) = \sum c_{j_1 \cdots j_p} u_1^{j_1} \cdots u_p^{j_p} = \sum_{(j)} c_j u^j, \tag{II.2.1}$$

that converges to $f(u)$ on the whole space $\mathbf{C}^p$. Then this series necessarily converges absolutely and uniformly over each p-disk $|u_j| < R_j < +\infty$, $j = 1, \ldots, p$. The partial derivatives of $f(u)$ of any order are also entire functions, and moreover their expansions

$$f(u) = \sum_{n=0}^{\infty} \frac{1}{n!} \left(\frac{\partial^n f}{\partial u_j^n} \right)_{u_j = 0} u_j^n \qquad (j = 1, \ldots, p),$$

exist and also converge absolutely and uniformly over any p-disk. The coefficients of the series (II.2.1) are equal to the values of the corresponding partial derivatives of $f(u)$ at the origin:

$$j_1! \cdots j_p! c_{j_1 \cdots j_p} = \left(\frac{\partial^{j_1 + \cdots + j_p} f}{\partial u_1^{j_1} \cdots \partial u_p^{j_p}} \right)_{u=0}$$

and can be estimated by Cauchy's inequality

$$|c_{j_1 \cdots j_p}| < \frac{M(R_1, \ldots, R_p)}{R_1^{j_1} \cdots R_p^{j_p}},$$

where $M(R_1, \ldots, R_p) = \max|f(u)|$ provided that $|u_j| = R_j$ $(j = 1, \ldots, p)$. **Liouville's theorem**, that *an entire function bounded in absolute value over the whole space* $\mathbf{C}^p$ *is a constant*, follows from this.

An equivalent definition of entire function can be stated as follows: $f(u)$ *is single-valued and continuous* on the space $\mathbf{C}^p$, with first order partial derivatives $\partial f / \partial u_j$ $(j = 1, \ldots, p)$ at each point. It follows immediately from this that if $f(u)$ is an entire function of p variables and $g(z)$ is an entire function of one variable, then $g[f(u)]$ is also an entire function of p variables. An example of such a function is $\exp[f(u)] = F(u)$. It does not have any zeros in $\mathbf{C}^p$. The converse is also true: if $g(u)$ is an entire function with no zeros in $\mathbf{C}^p$, then there exists an entire function $f(u)$ such that $g(u) = \exp f(u)$. Indeed, it suffices to take as $f(u)$ any of the single-valued continuous branches of $\operatorname{Ln}[g(u)]$ over $\mathbf{C}^p$.

A function $F(u)$ is called *meromorphic* if it can be expressed as the quotient of two entire functions

$$F(u) = \frac{\varphi(u)}{\psi(u)} \qquad (\psi(u) \not\equiv 0). \tag{II.2.2}$$

Of course, this representation is not unique since $\varphi(u)$ and $\psi(u)$ can be multiplied by any entire function $f(u)$.

A meromorphic function is analytic at all points where $\psi(u) \neq 0$. If $\psi(a) = 0$ and $\varphi(a) \neq 0$, then the point a is called a *pole* of the function $F(u)$. *When $p > 1$ the poles, as well as the zeros, are not isolated.* Indeed, for a pole there is a neighborhood in which $\varphi(u) \neq 0$. In this neighborhood, the manifold of zeros of the function $\psi(u)$, which as we noted at the end of section II.1 is a $(p-1)$-dimensional manifold, is at the same time the manifold of poles of the function $f(u)$.

Now let $\psi(a) = 0$ and $\varphi(a) = 0$. If the functions $\varphi(u)$ and $\psi(u)$ are not relatively prime, then let $\chi(u)$ be their *greatest common divisor* (which is defined as a divisor divisible by every common divisor of $\varphi(u)$ and $\psi(u)$). It can easily be found if we know the factorizations of $\varphi(u)$ and $\psi(u)$ into prime factors. The Euclidean algorithm can be used for the same purpose. Set

$$\varphi(u) = \chi(u)\varphi_1(u) \quad \text{and} \quad \psi(u) = \chi(u)\psi_1(u).$$

Then in a neighborhood of the point a we have $f(u) = \varphi_1(u)/\psi_1(u)$, where φ_1 and ψ_1 are analytic at a but not necessarily entire functions. Now there are three possibilities:

(α) $\psi_1(a) \neq 0$; then a is a regular point of $f(u)$, i.e., $f(u)$ is analytic at this point;

(β) $\psi_1(a) = 0$ but $\varphi_1(a) \neq 0$; then a is a singular point and a pole of $f(u)$;

(γ) $\psi_1(a) = 0$ and $\varphi_1(a) = 0$; then a is again a singular point of $f(u)$, called *a point of indeterminacy*. One can understand points of indeterminacy by looking at the simplest example, when $p = 2$ and $f(u)$ is a rational function of the form $f(u) = u_2/u_1$. Here the origin is a point of indeterminacy. If A is any complex number, then at all points of the one-dimensional complex manifold $u_2 - Au_1 = 0$, except at the origin, we have $f(u) = A$. This explains why singular points of this type are called points of indeterminacy.

Now we return to the general case and, for simplicity, let $a = 0$. Using the Weierstrass preparation theorem, we represent $\varphi_1(u)$ and $\psi_1(u)$ by formulas of the form (II.1.3):

$$\begin{aligned}\varphi_1(u) &= (\pi_0 u_1^{\nu_1} + \pi_1 u_1^{\nu_1 - 1} + \cdots + \pi_{\nu_1})\varphi_2(u), \\ \psi_1(u) &= (\rho_0 u_1^{\mu_1} + \cdots + \rho_{\mu_1})\psi_2(u),\end{aligned} \tag{II.2.3}$$

where $\varphi_2(u)$ and $\psi_2(u)$ are units at the point $u = 0$. Pseudopolynomials in these decompositions are relatively prime. It is possible to prove that they must also be relatively prime over a sufficiently small neighborhood of the point $u = 0$ (see Appendix B.2 at the end of the book). Let b be a point of this neighborhood at which $\varphi_1(b) = \psi_1(b) = 0$. Then b is also a point of indeterminacy of $f(u)$, by definition. In other words, there is a neighborhood of the point of indeterminacy $u = 0$ in which every point of the analytic manifold $\mathfrak{M}$ defined by the system

$$\mathfrak{M}\colon \pi_0 u_1^{\nu_1} + \pi_1 u_1^{\nu_1 - 1} + \cdots + \pi_{\nu_1} = 0, \qquad \rho_0 u_1^{\mu_1} + \cdots + \rho_{\mu_1} = 0, \tag{II.2.4}$$

is also a point of indeterminacy. (Here we assume that $p > 1$.) Since each of the two equations of the system defines a $(p-1)$-dimensional complex manifold, the system defines a $(p-2)$-dimensional manifold whose equation, if we eliminate u_1 from (II.2.4), can be given in the form

$$\Phi(u_2, \ldots, u_p) = 0. \tag{II.2.5}$$

When $p > 2$ $\mathfrak{M}$ is an analytic *hypersurface* in the space $\mathbf{C}^{p-1}$. Thus if we exclude the case $p = 1$ (when no points of indeterminacy exist), or when $p = 2$ (they are *isolated* and correspond to roots of the equation $\Phi(u_2) = 0$), *the points of indeterminacy, like the poles, are not isolated.* However, the complex dimension of the set of these points is one less than the dimension of the set of poles. For an arbitrary complex number A we now consider the $(p-1)$-dimensional manifold

$$\mathfrak{N}\colon (\pi_0 u_1^{\nu_1} + \cdots + \pi_{\nu_1}) - A\frac{\psi_2(0)}{\varphi_2(0)}(\rho_0 u_1^{\mu_1} + \cdots + \rho_{\mu_1}) = 0. \tag{II.2.6}$$

The point $u = 0$ belongs to $\mathfrak{N}$ (it is an interior point). It is obvious that if $u \in \mathfrak{N}$ and $\psi_1(u) = 0$, then $\rho_0 u_1^{\mu_1} + \cdots + \rho_{\mu_1} = 0$ (because of (II.2.3)), from which it follows that $u \in \mathfrak{M}$. But the dimension of $\mathfrak{M}$ is one less than the dimension of $\mathfrak{N}$. Hence in any neighborhood of the point $u = 0$ there are points belonging to $\mathfrak{N}$ which are not in $\mathfrak{M}$. At these points the condition $\psi_1(u) \neq 0$ must hold, i.e., these points are regular for $f(u)$. Taking the limit as $u \in \mathfrak{N}\backslash\mathfrak{M}$ approaches 0, we find, by (II.2.3) and (II.2.6), that

$$\lim_{u \in \mathfrak{N}\backslash\mathfrak{M},\, u \to 0} f(u) = \lim_{u \to 0} \frac{\varphi_1(u)}{\psi_1(u)} = A\frac{\psi_2(0)}{\varphi_2(0)} \cdot \frac{\varphi_2(0)}{\psi_2(0)} = A.$$

Thus for an arbitrary meromorphic function $f(u)$ the points of $\mathbf{C}^p$ fall into three categories.

(α) regular points, at which a finite limit of $f(u)$ exists;

(β) poles, at which the limit of $f(u)$ is ∞;

(γ) points of indeterminacy, at which every complex number is a limit point for the values of $f(u)$.

These criteria do not require special consideration of the functions $\varphi(u)$ and $\psi(u)$ separately in the representation (II.2.2).

The following fundamental theorem was given by Poincaré and Cousin (see Appendix B.3). *For each meromorphic function $f(u)$ there exist two entire functions $\varphi(u)$ and $\psi(u)$ such that $f(u) = \varphi(u)/\psi(u)$, and moreover at each point $u \in \mathbf{C}^p$ the functions $\varphi(u)$ and $\psi(u)$ are relatively prime.* A representation of this type will be called *reduced.*

Because of what was said above, we can assert that for any reduced representation the set of all singular points (i.e., poles as well as points of indeterminacy) coincides precisely with the $(p-1)$-dimensional manifold $\psi(u) = 0$, and all points of indeterminacy form a $(p-2)$-dimensional submanifold, defined by the two equations $\psi(u) = 0$ and $\varphi(u) = 0$. Of course, every function $f(u)$ has an infinite number of reduced representations. One reduced representation can be obtained from another by multiplying the numerator and the denominator of the quotient by any entire function that has no zeros, i.e., by a function of the type $\exp[g(z)]$, where $g(z)$ is an entire function. But *the converse is also true.* Let

$$f(u) = \frac{\varphi_1(u)}{\psi_1(u)} = \frac{\varphi_2(u)}{\psi_2(u)}$$

be two reduced representations of $f(u)$. Then $\varphi_1(u)\psi_2(u) = \varphi_2(u)\psi_1(u)$, i.e., $\varphi_1(u)|\varphi_2(u)\psi_1(u)$. But, by assumption, $\varphi_1(u)$ and $\psi_1(u)$ are relatively prime. Therefore $\varphi_1(u)|\varphi_2(u)$ at any point u, i.e., $\varphi_2(u) = \varphi_1(u)\alpha(u)$, where $\alpha(u)$ is an entire function. In exactly the same way, $\varphi_2(u)|\varphi_1(u)\psi_2(u)$ and since $\varphi_2(u)$ and $\psi_2(u)$ are relatively prime, $\varphi_1(u) = \varphi_2(u)\beta(u)$, where $\beta(u)$ is also an entire function. Multiplying, we get $\alpha(u)\beta(u) = 1$, i.e., $\alpha(u)$ does not have zeros and therefore $\alpha(u) = \exp[g(u)]$, i.e., $\varphi_2(u) = \varphi_1(u)\exp[g(u)]$. It is obvious that $\psi_2(u) = \psi_1(u)\exp[g(u)]$ also. Thus *if $f(u) = \varphi(u)/\psi(u)$ is some reduced representation of a meromorphic function, then any other reduced representation of the function has the form*

$$f(u) = \frac{\varphi(u)\exp g(u)}{\psi(u)\exp g(u)},$$

where $g(u)$ is an arbitrary entire function.

§3. The set of periods. Infinitesimal periods

If there is a p-dimensional vector $\omega = (\omega_1 \cdots \omega_p)^{\mathrm{T}}$, different from zero and such that for any meromorphic function $f(u)$ the following equality holds at all its regular points:

$$f(u+\omega) = f(u), \tag{II.3.1}$$

then the function is called *periodic* and ω is called its *period.*

We must keep in mind in everything that follows that a period of a function of p complex variables is an ordered collection of p complex numbers $\omega_1, \dots, \omega_p$. In other words, the equality (II.3.1) means that

$$f(u_1 + \omega_1, \dots, u_p + \omega_p) = f(u).$$

It follows from the definition that $u + \omega$ and u are regular points of a function $f(u)$ simultaneously, i.e., that *the set of all regular points is transformed into itself by a shift equal to a period.* Since the character of a singular point (pole or a point of indeterminacy) is completely determined by the behavior of the function at the regular points which lie in a neighborhood of this singular point, it follows that *the set of singular points is transformed into itself by a shift by a period* ω *and moreover that the set of poles is transformed into itself and the set of points of indeterminacy is transformed into itself as well.*

It is obvious that if ω is a period, then all vectors of the form $n\omega$, where n is any integer, are periods too. In order not to make $n = 0$ an exception we must agree that the zero vector is a period of any periodic function. It is obvious also that if $\omega^{(1)}, \dots, \omega^{(n)}$ are periods of a function $f(u)$ and $m_1, \dots, m_n$ are any integers, then

$$\omega = m_1\omega^{(1)} + \cdots + m_n\omega^{(n)}$$

is also a period of the same function.

We will frequently have to use a linear transformation of the variables, given by a formula

$$u^* = Au, \tag{II.3.2}$$

where $A = (a_{jk})$, $j = 1, \dots, p$, $k = 1, \dots, p$ is a square p by p matrix with a nonzero determinant $|A|$. This transformation, of course, transforms entire functions into entire functions, meromorphic functions into meromorphic functions, and their reduced representations into reduced representations.

Let us investigate what happens to periods of $f(u)$ under these circumstances. Set

$$f(u) = f(A^{-1}u^*) = f^*(u^*). \tag{II.3.3}$$

Assume that ω^* is a period of $f^*(u^*)$. Then $f^*(u^* + \omega^*) = f^*(\omega^*)$ at each regular point of the function $f^*(\omega^*)$; this means that

$$f(A^{-1}u^* + A^{-1}\omega^*) = f(A^{-1}u^*), \quad \text{i.e.,}\ f(u + A^{-1}\omega^*) = f(u).$$

Thus, $A^{-1}\omega^* = \omega$ is a period of $f(u)$, i.e.,

$$\omega^* = A\omega. \tag{II.3.4}$$

It is easy to see that the converse is also true: for each period ω of $f(u)$ this formula gives a corresponding period ω^* of the function $f^*(\omega^*)$. Therefore, *when a linear transformation acts on the variables, the same linear transformation acts on the periods of the function.*

It is obvious that the set Π of periods of any periodic function is an infinite unbounded set. There are two possibilities here: (α) Π has finite limit points or at least one such point; (β) Π does not have any finite limit points, i.e., each point of the set of periods is an isolated point.

Let us study the case (α) in detail. Let $\omega^{(0)}$ be a limit point (vector) of the set of periods and let $\{\omega^{(n)}\}$ be a sequence of pairwise distinct periods converging to $\omega^{(0)}$. For any regular point $u^{(0)}$ of $f(u)$ we have

$$f(u^{(0)} + \omega^{(n)}) = f(u^{(0)}). \tag{II.3.5}$$

If $|u_j - u_j^{(0)}| < \rho_j$, $j = 1, \dots, p$, is a p-disk which contains no singular points of $f(u)$, then the p-disk $|u_j - (u_j^{(0)} + \omega_j^{(n)})| < \rho_j$, $j = 1, \dots, p$, also has no singular points of $f(u)$. Since $\lim_{n\to\infty}(u_j^{(0)} + \omega_j^{(n)}) = u_j^{(0)} + \omega_j^{(0)}$, it follows that the limiting p-disk $|u_j - (u_j^{(0)} + \omega_j^{(0)})| < \rho_j$, $(j = 1, \dots, p)$ does not contain any singular points, i.e., the point $u^{(0)} + \omega^{(0)}$ is also regular. Taking the limit in (III.3.5) we find that for any regular point $u^{(0)}$

$$f(u^{(0)} + \omega^{(0)}) = f(u^{(0)}),$$

i.e., $\omega^{(0)}$ is also a period. Hence Π is a *closed set* (in the case (β) this is trivial). But since $\omega^{(0)} = \lim_{n\to\infty} \omega^{(n)}$ is a period it follows that $\{\omega^{(n)} - \omega^{(0)}\}$ is a sequence of nonzero periods converging to zero. We will call the terms of such a sequence *infinitesimal periods*. Thus, case (α) is possible if and only if $f(u)$ allows infinitesimal periods. In this case, each period $\omega \in \Pi$ will be a limit point of Π since $\omega = \lim_{n\to\infty}[\omega + (\omega^{(0)} - \omega^{(n)})]$. Therefore, in case (α), the set of periods is perfect.

As we have seen, the set Π of periods is transformed, under a linear transformation of the variables $u^* = Au$, by the same transformation $\Pi^* = A\Pi$. Because of the continuity of the transformation it follows that infinitesimal periods (if they exist) are transformed into infinitesimal periods.

We call a meromorphic function $f(u)$ of p complex variables *degenerate* if it actually depends on fewer than p of the variables. For example, each meromorphic function of one variable is degenerate if it is considered as a function of more variables.

We show that *in order for $f(u)$ to be degenerate it is necessary and sufficient that there exist a p-dimensional vector $\gamma = (\gamma_1 \cdots \gamma_p)^{\mathrm{T}}$, not equal to zero and such that*

$$\gamma_1 \frac{\partial f(u)}{\partial u_1} + \cdots + \gamma_p \frac{\partial f(u)}{\partial u_p} = 0 \tag{II.3.6}$$

holds at each regular point of $f(u)$. In other words, the scalar product of $\operatorname{grad} f$ by some *constant vector* γ $(\gamma \neq 0)$ must be equal to zero at a regular point of $f(u)$.

Indeed, if $f(u)$ does not depend on, for example, u_p then $\partial f/\partial u_p = 0$ and (II.3.6) holds when $\gamma_1 = \cdots = \gamma_{p-1} = 0$, $\gamma_p = 1$. Also, if the function

$f^*(u^*)$ obtained by the transformation $u = Bu^*$ does not depend on u_p^*, then

$$\frac{\partial f^*(u^*)}{\partial u_p^*} = \gamma_1 \frac{\partial f}{\partial u_1} + \cdots + \gamma_p \frac{\partial f}{\partial u_p} = 0,$$

where $\gamma_1, \ldots, \gamma_p$, in that order, are the elements of the last column of the matrix B. Since, by definition, $|B| \neq 0$, not all the numbers γ_j ($j = 1, \ldots, p$) are equal to zero. Thus the condition (II.3.6) is necessary for $f(u)$ to be a degenerate function. It is also sufficient. Indeed, assuming that this condition holds, we construct a square matrix B of order (p, p) whose last column consists of the numbers γ_j ($j = 1, \ldots, p$) and whose determinant $|B|$ is not zero. Then, after a change of variables $u = Bu^*$, we find that

$$\frac{\partial f^*(u^*)}{\partial u_p^*} = \gamma_1 \frac{\partial f}{\partial u_1} + \cdots + \gamma_p \frac{\partial f}{\partial u_p} = 0,$$

i.e. $f^*(u^*)$ does not depend on u_p^* and therefore $f(u)$ is a degenerate function.

It is now easy to prove the following important theorem.

THEOREM II.3.1. *For a meromorphic function $f(u)$ of p complex variables to be degenerate, it is necessary and sufficient that this function admits infinitesimal periods.*

Since a linear change of variables maps infinitesimal periods into infinitesimal periods, it follows that the criterion of the theorem is indeed invariant under any nonsingular linear transformation. We also note that in this proof we will be directly using only the fact that $f(u)$ has a total differential (at each regular point). Hence the theorem is also true for nonanalytic functions of p real variables if these functions are collectively differentiable in their variables over some open subset of the space $\mathbf{R}^p$, where this set must be mapped into itself by a shift equal to any period.

The necessity of the condition of the theorem is obvious. Let $f(u)$ be a degenerate function. Assume that the appropriate linear transformation has already been applied and that $f(u)$ does not depend on u_p. Then for any complex number c we have

$$f(u_1, \ldots, u_p + c) = f(u_1, \ldots, u_p).$$

But this means that the function has a period all of whose components except perhaps the last are equal to zero. Setting, for example, $c = 1/n$, $n = 1, 2, \ldots$, we see that $f(u)$ admits infinitesimal periods.

We now prove the sufficiency. Let $\{\omega^{(n)}\}$ be a sequence of nonzero periods of $f(u)$, converging to zero. Choose p different regular points of $f(u)$: $c^{(1)}, \ldots, c^{(p)}$ (the points at which the function is differentiable in the totality of its variables). Denote, for brevity, the derivatives $\partial f/\partial u_j$ at the point

$u = c^{(k)}$ by $\partial f_k/\partial u_j$ and let $\omega_j^{(n)}$ $(j = 1, \dots, p)$ be the components of the vector $\omega^{(n)}$. Then

$$f(c^{(k)} + \omega^{(n)}) - f(c^{(k)}) = f(c_1^{(k)} + \omega_1^{(n)}, \dots, c_p^{(k)} + \omega_p^{(n)}) - f(c^{(k)})$$
$$= \sum_{j=1}^{p} \left(\frac{\partial f_k}{\partial u_j} + \varepsilon_{k,j}^{(n)} \right) \omega_j^{(n)} = 0,$$

where $\lim_{n\to\infty} \varepsilon_{k,j}^{(n)} = 0$ for all $k = 1, \dots, p$, $j = 1, \dots, p$. It follows from this that for each n the determinant of the latter system of equations $(k = 1, \dots, p)$ is equal to zero:

$$\left| \frac{\partial f_k}{\partial u_j} + \varepsilon_{k,j}^{(n)} \right| = 0,$$

or that the following determinant:

$$\left| \frac{\partial f_k}{\partial u_j} \right| = 0 \tag{II.3.7}$$

is equal to zero when one takes the limit as $n \to \infty$. Each row of this determinant consists of coordinates of the vector $\operatorname{grad} f(u)$ at a corresponding point $u = c^{(k)}$. The equality (II.3.7) means that for any p regular points, the p gradients of f computed at these points are linearly dependent. If $f(u) \not\equiv \text{const}$ (otherwise the theorem is trivial) there exists a point at which $\operatorname{grad} f \neq 0$. We denote by s the maximum number of regular points for which the s vectors $\operatorname{grad} f$ are linearly independent. It is obvious that $1 \leq s < p$. Let $c^{(1)}, \dots, c^{(s)}$ be such points. Denote by $\mathscr{L}$ the linear manifold spanned by the vectors $\operatorname{grad} f(c^{(1)}), \dots, \operatorname{grad} f(c^{(s)})$. For each regular point u the gradient of f belongs to $\mathscr{L}$. Since the dimension of $\mathscr{L}$ is less than p there is a nonzero vector γ orthogonal to $\mathscr{L}$ and therefore orthogonal to $\operatorname{grad}[f(u)]$:

$$(\operatorname{grad} f, \gamma) = \gamma_1 \frac{\partial f(u)}{\partial u_1} + \cdots + \gamma_p \frac{\partial f(u)}{\partial u_p} = 0.$$

Thus, $f(u)$ is a degenerate function.

We will assume in what follows (unless the contrary is stated) that $f(u)$ is not a degenerate function. Hence the set of all periods of the function $f(u)$ will have no finite limit points.

§4. Conditions for independence of periods

The vectors $\omega^{(1)}, \dots, \omega^{(n)}$ in the space $\mathbf{C}^p$ are said to be *independent over integers* or just *independent* if it follows that $m_1 = \cdots = m_n = 0$ when

$$m_1\omega^{(1)} + \cdots + m_n\omega^{(n)} = 0, \tag{II.4.1}$$

where m_j $(j = 1, \dots, n)$ are integers. If we forget for a moment that the discussion will concern the periods of nondegenerate functions of p complex variables, then arbitrarily many vectors $\omega^{(j)}$ independent in the narrow sense just mentioned can be constructed for a given p. For example, let α be a transcendental or algebraic number of degree not less than n (i.e., such that it does not satisfy any algebraic equation of degree less than n with integer coefficients). Then from the equality $m_1 + m_2\alpha + \cdots + m_n\alpha^{n-1} = 0$, where m_j are integers, it necessarily follows that $m_1 = m_2 = \cdots = m_n = 0$. Assuming that the first coordinate of the p-dimensional vector $\omega^{(j)}$ is α^{j-1} $(j = 1, \ldots, n)$, and that the rest of the coordinates are any complex numbers we obtain a simple example of n independent vectors in $\mathbf{C}^p$ for any natural n.

A different behavior is displayed by the condition of *real independence*: vectors $\omega^{(1)}, \ldots, \omega^{(n)}$ in the space $\mathbf{C}^p$ are called *independent over the reals* if it follows from a relation

$$x_1\omega^{(1)} + \cdots + x_n\omega^{(n)} = 0, \tag{II.4.2}$$

where x_j $(j = 1, \ldots, n)$ are real numbers, that $x_1 = x_2 = \cdots = x_n = 0$.

It is not hard to show that *the maximal number of real independent vectors in* $\mathbf{C}^p$ is equal to $2p$, i.e., $n \leq 2p$ is always true. Indeed, the coordinates $\omega_k^{(j)}$ $(k = 1, \ldots, p)$ of the vector $\omega^{(j)}$ are complex numbers $\omega_k^{(j)} = \alpha_k^{(j)} + i\beta_k^{(j)}$. Hence each vector $\omega^{(j)}$ in $\mathbf{C}^p$ can be considered as a vector of the $2p$-dimensional real space $\mathbf{R}^{2p}$ with coordinates $\alpha_1^{(j)}, \ldots, \alpha_p^{(j)}, \beta_1^{(j)}, \ldots, \beta_p^{(j)}$. From this point of view the condition of real linear independence of $\omega^{(j)}$ in $\mathbf{C}^p$ expresses the linear independence of corresponding $2p$-dimensional vectors in $\mathbf{R}^{2p}$. But the maximum number of such vectors is equal to $2p$.

We will now prove that if $\omega^{(j)}$ are periods of a nondegenerate function $f(u)$ of complex variables, then the conditions (II.4.1) and (II.4.2) are equivalent for them.

THEOREM II.4.1. *Independent periods* $\omega^{(1)}, \ldots, \omega^{(n)}$ *of a nondegenerate function* $f(u)$ *of* p *complex variables are at the same time real independent* (*the converse is trivial*).

Assume that $\omega^{(1)}, \ldots, \omega^{(n)}$ are independent but for some real numbers $x_1, \ldots, x_n$

$$x_1\omega^{(1)} + \cdots + x_n\omega^{(n)} = 0. \tag{II.4.3}$$

We will prove that then $x_1 = \cdots = x_n = 0$. For each natural number ν we select n integers $m_{\nu,1}, \ldots, m_{\nu,n}$ such that the inequalities

$$|\nu x_j - m_{\nu,j}| \leq \frac{1}{2}, \qquad j = 1, \ldots, n, \nu = 1, 2, \ldots \tag{II.4.4}$$

hold. Then the vectors $\Omega^{(\nu)}$, defined by the formulas

$$\Omega^{(\nu)} = -\sum_{j=1}^{n} m_{\nu,j}\omega^{(j)} = \sum_{j=1}^{n}(\nu x_j - m_{\nu,j})\omega^{(j)}, \qquad \nu = 1, 2, \ldots \tag{II.4.5}$$

will also be periods of the function $f(u)$. Because of (II.4.4) the sequence $\{\Omega^{(\nu)}\}$ is bounded:

$$\|\Omega^{(\nu)}\| \leq \frac{1}{2}(\|\omega^{(1)}\| + \cdots + \|\omega^{(n)}\|). \tag{II.4.5$'$}$$

Since the set of periods of the function $f(u)$ does not have finite limit points, there can be only a finite number of distinct vectors among the terms of the sequence (II.4.5). Let the vector α occur an infinite number of times in this sequence:

$$\Omega^{(\nu_0)} = \Omega^{(\nu_1)} = \cdots = \alpha, \qquad \nu_0 < \nu_1 < \cdots \tag{II.4.6}$$

Then

$$\Omega^{(\nu_k)} - \Omega^{(\nu_0)} = \sum_{j=1}^{n}[(\nu_k - \nu_0)x_j - (m_{\nu_k,j} - m_{\nu_0,j})]\omega^{(j)} = 0 \tag{II.4.7}$$

if $k = 1, 2, \ldots$. Because of the inequalities (II.4.4) we have

$$|(\nu_k - \nu_0)x_j - (m_{\nu_k,j} - m_{\nu_0,j})| \leq 1. \tag{II.4.8}$$

On the other hand, in view of (II.4.3), the equalities (II.4.7) can be rewritten in the following form:

$$\sum_{j=1}^{n}(m_{\nu_k,j} - m_{\nu_0,j})\omega^{(j)} = 0. \tag{II.4.9}$$

But the periods $\omega^{(j)}$ are by definition independent. Hence it follows from (II.4.9) that $m_{\nu_k,j} - m_{\nu_0,j} = 0$, $j = 1, \ldots, n$; $k = 1, 2, \ldots,$ and therefore (by the inequalities (II.4.8)),

$$|(\nu_k - \nu_0)x_j| \leq 1 \qquad (j = 1, \ldots, n;\ k = 1, 2, \ldots).$$

Here $\nu_k \to \infty$ when $k \to \infty$ from which we get $x_1 = \cdots = x_n = 0$. This completes the proof of the theorem.

Since we now know that the independent periods of a nondegenerate function of p complex variables are also linearly independent over the reals, we obtain the *Riemann-Weierstrass Theorem* as a corollary of Theorem II.4.1: *the number of independent periods of a nondegenerate function of p complex variables cannot exceed $2p$.*

Another kind of linear independence of the periods plays a significant role in our discussion. We say that the periods $\omega^{(1)}, \ldots, \omega^{(n)}$ are complex independent if it follows from the relation

$$\lambda_1\omega^{(1)} + \cdots + \lambda_n\omega^{(n)} = 0, \tag{II.4.10}$$

(where λ_j are complex numbers) that $\lambda_1 = \cdots = \lambda_n = 0$. Since $\omega^{(j)}$ are vectors of the space $\mathbf{C}^p$, the largest possible number of complex independent periods is equal to p. For example, the vectors $\omega^{(j)}$, all of whose coordinates are zero except the jth coordinate which is equal to 1 $(j = 1, \dots, p)$, are complex independent. The vectors

$$\alpha = \begin{pmatrix} 1 \\ 0 \\ \vdots \\ 0 \end{pmatrix} \quad \text{and} \quad \beta = \begin{pmatrix} i \\ 0 \\ \vdots \\ 0 \end{pmatrix}$$

are real independent but they are not complex independent, because of the relation $\alpha i - \beta = 0$.

We will show by a simple example that there are nondegenerate meromorphic functions of p complex variables with $2p$ independent periods. It would be easiest to take the elliptic function $\wp(z)$ of Weierstrass as a basic building block. However, in order to use the facts given in Chapter I, we will base our discussion on the functions $\alpha(z) = \vartheta_3(z)/\vartheta_4(z)$, where $\vartheta_3(z)$ and $\vartheta_4(z)$ are theta functions (section I.3). This quotient is a meromorphic function since $\vartheta_3(z)$ and $\vartheta_4(z)$ are entire functions. It has two independent periods, 1 and 2τ $(\operatorname{Im}\tau > 0)$. Moreover, $\alpha(z)$ is nondegenerate, i.e., it is not equal to a constant. Indeed, from Table 1 (section I.3) and the series defining the theta functions we obtain

$$\alpha\left(\frac{\tau}{2}\right) = -i\frac{\vartheta_2}{\vartheta_1} = \infty$$

$(\vartheta_1 = \vartheta_1(0) = 0,\ \vartheta_2 = \vartheta_2(0) = 2q^{1/4} + 2q^{9/4} + \cdots \neq 0$ because $|q| < 1)$ and $\alpha(1/2 + \tau/2) = i\vartheta_1/\vartheta_2 = 0$. We note that it follows, taking into account the fact that $\tau/2$ is a pole of $\alpha(z)$, that $\alpha'(\tau/2) = \infty$. Now set $f(u) = \sum_{j=1}^{p} c_j\alpha(u_j)$, where $c_j \neq 0$ $(j = 1, \dots, p)$. It is obvious that $2p$ vectors $\pi^{(1)}, \omega^{(1)}, \pi^{(2)}, \omega^{(2)}, \dots, \pi^{(p)}, \omega^{(p)}$, where all the coordinates of the vectors $\pi^{(k)}$ and $\omega^{(k)}$ except the kth are equal to zero and the kth coordinate is equal to 1 for $\pi^{(k)}$ and 2τ for $\omega^{(k)}$, are periods and in fact independent periods of $f(u)$. Indeed, because of the definition and the properties of theta functions, we have

$$f(u + \pi^{(k)}) = c_k\alpha(u_k + 1) + \sum_{j\neq k} c_j\alpha(u_j) = f(u)$$

and analogously $f(u + \omega^{(k)}) = f(u)$. Among the periods mentioned, the p vectors $\pi^{(j)}$ as well as the p vectors $\omega^{(k)}$ are complex independent.

It remains to check that $f(u)$ is a nondegenerate function. The criterion expressed by the equation (II.3.6) will be used for this purpose. In the case under consideration when $\partial f/\partial u_j = c_j\alpha'(u_j)$, the corresponding equation has the form

$$\sum_{j=1}^{p} \gamma_j c_j \alpha'(u_j) = 0.$$

Let a be a regular point of the function $\alpha(z)$. We fix k with $1 \leq k \leq p$ and set $u_j = a$ for each $j \neq k$. We let u_k approach (with no restrictions) the point $\tau/2$, which is a pole of $\alpha'(z)$ as was noted above. Then it follows from the latter equation that $\gamma_k c_k = 0$; and since $c_k \neq 0$, we have $\gamma_k = 0$. This is true for all k, $1 \leq k \leq p$, and therefore $f(u)$ is nondegenerate.

In this example the function $f(u)$ is meromorphic, although it is of a quite special type (it does not have points of indeterminacy at all). It would be fruitless to search for an entire function of p variables with $2p$ independent periods. We will show later (section II.7) that there is no entire function of p variables which is nondegenerate and whose number of periods exceeds p. But an entire function with p independent periods is easy to construct; an example is the function

$$g(u) = \sum_{j=1}^{p} C_j \exp u_j, \qquad C_j \neq 0 \ (j = 1, \dots, p).$$

Here each vector $\omega^{(k)}$ $(k = 1, \dots, p)$ is a period for which all the coordinates $\omega^{(k)}$ except the kth are zero and the kth is equal to $2\pi i$. It is obvious that the system $\omega^{(k)}$ is independent. Since $\partial f/\partial u_j = C_j \exp(u_j)$, the criterion (II.3.6) takes the form

$$\sum_{j=1}^{p} \gamma_j C_j \exp u_j = 0.$$

Here if we set $u_j = 0$ when $j \neq k$ and let u_k take unbounded real positive values, we obtain $\gamma_k = 0$, $k = 1, \dots, p$, from which it follows that $f(u)$ is a nondegenerate function.

§5. The fundamental system of periods

Let $f(u)$ be a nondegenerate periodic meromorphic function of p variables. We saw that among its periods no more than $2p$ are independent. Let r $(1 \leq r \leq 2p)$ denote the *maximal* number of independent periods of a given function. We consider a system of r independent periods $\omega^{(1)}, \dots, \omega^{(r)}$ of this function. If we add to them an arbitrary period ω, then the system so enlarged will not be independent and therefore there exist numbers $m_1, \dots, m_r$, m, not all zero, such that

$$m_1\omega^{(1)} + \dots + m_r\omega^{(r)} + m\omega = 0.$$

It is obvious that $m \neq 0$ (otherwise $\omega^{(1)}, \dots, \omega^{(r)}$ would not be independent) and hence

$$\omega = -\frac{m_1}{m}\omega^{(1)} - \dots - \frac{m_r}{m}\omega^{(r)}.$$

Thus any period of $f(u)$ can be expressed as a linear combination with rational coefficients of the periods in a maximal independent system. Among such systems we distinguish those for which the coefficients in the previous expression are *integers* (for any ω).

A system of independent periods $\omega^{(1)}, \dots, \omega^{(r)}$ of a nondegenerate function $f(u)$ is called *fundamental* (or *basic* or *primitive*) if every period ω of the function can be expressed in the form

$$\omega = m_1\omega^{(1)} + \cdots + m_r\omega^{(r)}, \qquad \text{(II.5.1)}$$

where $m_1, \dots, m_r$ are integers. Because the system $\omega^{(j)}$ is independent, such a representation is unique. It follows from the definition of a fundamental system that the set Π of all periods is countable. Hence the assumption that the function $f(u)$ is nondegenerate can be omitted from this definition. For, if a function admits infinitesimally small periods, then the set of its periods is perfect and is therefore uncountable (section II.3).

We will prove a simple theorem which contains another characteristic of a fundamental system. Then we will present a method of construction (and at the same time a proof of the existence) of a fundamental system. We first agree that in this section a period will be considered as a real vector in the space $\mathbf{R}^{2p}$.

THEOREM II.5.1. *Let $f(u)$ be a nondegenerate periodic function of p complex variables and let $\omega^{(1)}, \dots, \omega^{(r)}$ be an independent system of its periods $(r \le 2p)$ such that the system has the following properties*:

(a) *there are no nonzero periods of $f(u)$ among the points of the r-dimensional parallelotope P_r*

$$P_r\colon u = x_1\omega^{(1)} + \cdots + x_r\omega^{(r)}, \qquad 0 \le x_j < 1,\ j = 1, \dots, r,$$

spanned by the vectors $\omega^{(j)}$ $(j = 1, \dots, r)$;

(b) *every period of $f(u)$ belongs to a subspace $L_r \subset \mathbf{R}^{2p}$, spanned by the system $\omega^{(j)}$.*

Then $\omega^{(1)}, \dots, \omega^{(r)}$ is a fundamental system.

Indeed, any point $u \in L_r$ can be expressed uniquely in the form

$$u = y_1\omega^{(1)} + \cdots + y_r\omega^{(r)}$$

where y_j $(j = 1, \dots, r)$ are real numbers. Let ω be any period of $f(u)$. Because of condition (b) we have

$$\omega = a_1\omega^{(1)} + \cdots + a_r\omega^{(r)}. \qquad \text{(II.5.2)}$$

Represent each a_j in the form $a_j = \nu_j + \mu_j$ where ν_j is the integer part of a_j and $0 \le \mu_j < 1$. Then

$$\mu_1\omega^{(1)} + \cdots + \mu_r\omega^{(r)} = \omega - (\nu_1\omega^{(1)} + \cdots + \nu_r\omega^{(r)})$$

is also a period of $f(u)$ lying in P_r. Because of condition (a) this period must be zero. Hence $\mu_1 = \cdots = \mu_r = 0$ and all the coefficients in the expression (II.5.2) are integers. Thus $\omega^{(j)}$, $j = 1, \dots, r$, is a fundamental system.

The reader can easily check that the converse is also true: *if* $\omega^{(1)}, \dots, \omega^{(r)}$ *is a fundamental system, then conditions* (a) *and* (b) *hold for this system.*

We note that when $r = 2p$ condition (b) becomes trivial ($L_R = \mathbf{R}^{2p}$). Only condition (a) of Theorem II.5.1 then remains as a characteristic property of a fundamental system.

We now use induction to *construct the fundamental system for any nondegenerate periodic function of* p *complex variables.* To do this, we will show how one can move from some system $\omega^{(1)}, \dots, \omega^{(q)}$ for which condition (a) holds but (b) does not to a system $\omega^{(1)}, \dots, \omega^{(q)}, \omega^{(q+1)}$ for which condition (a) still holds. If (b) does not hold for $q \le 2p - 1$, then it must hold for $q = 2p$. In order to take the first step we consider some period $\omega \neq 0$ and with it all vectors of the form $y\omega$, where y is a real number. If there are no nonzero periods among the vectors $x\omega$ for which $0 \le x < 1$, we set $\omega^{(1)} = \omega$. This is the starting point of our construction. If, on the other hand, such periods exist (i.e., condition (a) does not hold for $r = 1$), then there are only finitely many such periods (infinitesimal because periods do not exist!); we denote by $\omega^{(1)}$ the period for which x has the smallest positive value. Assume that a system of independent periods $\omega^{(1)}, \dots, \omega^{(q)}$ $(1 \le q \le 2p - 1)$ for which condition (a) holds has been constructed. If condition (b) holds, this is indeed the fundamental system we are looking for. Otherwise there exists a period ω which does not belong to the subspace determined by the vectors $\omega^{(j)}$ $(j = 1, \dots, q)$. In other words, ω cannot be expressed in the form $y_1\omega^{(1)} + \dots + y_q\omega^{(q)}$ where y_j are real numbers. We then adjoin ω to the system $\omega^{(j)}$ and consider a subspace $\mathscr{L}'$ determined by the extended system of vectors. Let P' be the parallelotope built on the vectors $\omega^{(1)}, \dots, \omega^{(q)}, \omega$:

$$u = x_1\omega^{(1)} + \dots + x_q\omega^{(q)} + x\omega,$$
$$0 \le x_1 < 1, \dots, 0 \le x_q < 1, \qquad 0 \le x < 1.$$

If condition (a) holds for P', we set $\omega^{(q+1)} = \omega$ and consider the resulting system of $q+1$ independent periods of $f(u)$. If condition (a) does not hold, then because only a finite number of distinct periods can exist in P' we find a period whose x coordinate has the least positive value. Let this coordinate be ξ and denote one (if there are more than one) such period by $\omega^{(q+1)}$:

$$\omega^{(q+1)} = \xi_1\omega^{(1)} + \dots + \xi_q\omega^{(q)} + \xi_{q+1}\omega; \tag{II.5.3}$$

here $0 \le \xi_j < 1$ $(j = 1, \dots, q)$ and $\zeta_{q+1} = \xi$, $0 < \xi < 1$.

We now verify that condition (a) holds for the system $\omega^{(1)}, \dots, \omega^{(q)}$, $\omega^{(q+1)}$. Assume the contrary. Then there is a period $\omega^{(0)} \neq 0$ of the form

$\omega^{(0)} = \eta_1\omega^{(1)} + \cdots + \eta_q\omega^{(q)} + \eta_{q+1}\omega^{(q+1)}$, where $0 \le \eta_j < 1 (j = 1, \dots, q+1)$. It is obvious that $\eta_{(q+1)} \neq 0$ (otherwise condition (a) would be contradicted for the system $\omega^{(1)}, \dots, \omega^{(q)}$). If we replace $\omega^{(q+1)}$ by its expression (II.5.3), we obtain

$$\omega^{(0)} = (\eta_1 + \xi_1\eta_{q+1})\omega^{(1)} + \cdots + (\eta_q + \xi_q\eta_{q+1})\omega^{(q)} + \xi_{q+1}\eta_{q+1}\omega.$$

Here the numbers in the parentheses can be equal to 1 or or greater than 1 (but they all are less than 2). If we substitute for each $\eta_j + \xi_j\eta_{q+1}$ its fractional part α_j we change $\omega^{(0)}$ at most by a period and therefore obtain a new period

$$\omega^{(0)'} = \alpha_1\omega^{(1)} + \cdots + \alpha_q\omega^{(q)} + \xi_{q+1}\eta_{q+1}\omega; \tag{II.5.4}$$

here $0 \le \alpha_j < 1$ $(j = 1, \dots, q)$ and $0 < \xi_{q+1}\eta_{q+1} < \xi_{q+1} = \xi$. Comparing this with (II.5.3) we see that we have a contradiction of the assumption about the minimality of ξ. Thus a fundamental system will be constructed after a finite number of steps (at most $2p$).

We distinguish the case (the study of which will be our main concern in later chapters) when $r = 2p$, i.e., when a nondegenerate function $f(u)$ of p complex variables has the *maximum number of independent periods.* Let $\omega^{(1)}, \dots, \omega^{(2p)}$ be a fundamental system of its periods. Then the parallelotope Π_0, each point $u(0)$ of which can be expressed in the form

$$u^{(0)} = \alpha_1\omega^{(1)} + \cdots + \alpha_{2p}\omega^{(2p)}, \qquad 0 \le \alpha_j < 1, \\ j = 1, \dots, 2p, \tag{II.5.5}$$

will be a bounded subset of $\mathbf{C}^p$. This subset has complex dimension p (or real dimension $2p$). It satisfies the following conditions:

(A) Π_0 does not contain any nonzero periods of $f(u)$;

(B) for any point $u \in \mathbf{C}^p$ there is one and only one *congruent* point $u^{(0)} \in \Pi_0$ such that

$$u = u^{(0)} + m_1\omega^{(1)} + \cdots + m_{2p}\omega^{(2p)},$$

where the numbers $m_1, \dots, m_{2p}$ are integers.

Such a parallelotope will be called *the fundamental parallelotope* (or the fundamental region) *of* $f(u)$. We note that Π_0 can be considered as a *closed manifold without boundary* if the limit points for which one or more of the numbers α_j in (II.5.5) become 1 are identified with the congruent points of Π_0 (the corresponding 1's become 0's). This manifold is called a *torus* of complex dimension p. In the special case $p = 1$, Π_0 is an ordinary torus.

It follows from the definition of the fundamental parallelotope that the totality of values taken by $f(u)$ on $\mathbf{C}^p$ coincides with the totality of values taken by it on Π_0.

§6. Transformation of period matrix

Each fundamental system $\omega^{(1)}, \dots, \omega^{(r)}$ of periods can be considered either as a matrix with one row and r columns (then its elements are periods) or as a matrix with p rows and r columns (then its elements are complex numbers). Set

$$\Omega = (\omega^{(1)}, \dots, \omega^{(r)}) = \begin{pmatrix} \omega_1^{(1)} & \cdots & \omega_1^{(r)} \\ \cdots & \cdots & \cdots \\ \omega_p^{(1)} & \cdots & \omega_p^{(r)} \end{pmatrix}. \tag{II.6.1}$$

We will call such a matrix *the matrix of fundamental periods* or (more concisely) *the matrix of periods.* Let us find the relation between the matrices Π and Ω of $f(u)$. Let

$$\Pi = (\pi^{(1)}, \dots, \pi^{(\rho)}) = \begin{pmatrix} \pi_1^{(1)} & \cdots & \pi_1^{(\rho)} \\ \cdots & \cdots & \cdots \\ \pi_p^{(1)} & \cdots & \pi_p^{(\rho)} \end{pmatrix}. \tag{II.6.2}$$

Because the system of vectors $\omega^{(j)}$ must belong to a subspace spanned by the vectors of the system $\pi^{(k)}$, and conversely, it follows that $\rho = r$. Furthermore, each vector $\pi^{(j)}$ is (because the system Ω is fundamental) an integral linear combination of the vectors $\omega^{(k)}$:

$$\pi^{(j)} = m_{1j}\omega^{(1)} + \cdots + m_{rj}\omega^{(r)}, \qquad j = 1, \dots, \rho = r.$$

We denote by M a square matrix of coefficients

$$M = (m_{kj}), \qquad k = 1, \dots, r;\ j = 1, \dots, r. \tag{II.6.3}$$

We obtain the following relation between the $1 \times r$ matrices Π and Ω:

$$\Pi = \Omega M. \tag{II.6.4}$$

From now on, we will call an $n \times m$ matrix an (n, m) matrix. We can interchange the roles of Π and Ω in this argument. Then we have

$$\Omega = \Pi N, \tag{II.6.5}$$

where $N = (n_{kj})$ is a square matrix whose elements are integers. From (II.6.5) and (II.6.4) it follows that

$$\Omega = \Omega M N, \qquad \Omega(E_r - MN) = 0,$$

where E_r is a unit (r, r) matrix. All elements of the matrix $E_r - MN$ are integers, and the elements of the $(1, r)$ matrix $\Omega(E_r - MN)$ are vectors whose integral linear combinations are equal to zero. Because of the independence of the periods, all the coefficients of these combinations are also zero. Therefore

$$E_r - MN = 0, \qquad MN = E_r, \qquad N = M^{-1}. \tag{II.6.6}$$

Taking determinants, we find that $|M||N| = 1$, and since $|M|, |N|$ are integers

$$|M| = \pm 1. \tag{II.6.7}$$

A matrix whose elements are integers is called *unimodular* if its determinant is equal to ± 1.

Thus, *the transformation from one period matrix to another expressed by the relation* (II.6.4) *is a unimodular linear transformation.* It is easy to check that the converse is also true: *if a matrix is obtained from a fundamental matrix* Ω *by means of a unimodular transformation* (II.6.4), *then it is fundamental.*

Set $\omega_k^{(j)} = \sigma_k^{(j)} + i\tau_k^{(j)}$, where σ and τ are real numbers. Then Ω can be written in the form $\Omega = \Sigma + iT$, where $\Sigma = (\sigma_k^{(j)})$, $T = (\tau_k^{(j)})$, $j = 1, \dots, r$, $k = 1, \dots, p$. Form a $(2p, r)$ matrix $\binom{\Sigma}{T}$. Each of its r columns represents a period $\omega^{(j)}$ considered as a vector of the space $\mathbf{R}^{2p}$. Because $\omega^{(j)}$, $j = 1, \dots, r$, are real independent, the rank of this matrix must be r (recall that $r \le 2p$ always). However, the rank of the period matrix Ω can be less than r, although it is obviously not less than 1. Here the essential point is again the difference between the notions of *real* and *complex* independence. Let, for example, $p \ge 2$ and $r = 2$, and set

$$\Omega = \begin{pmatrix} 1 & i \\ 0 & 0 \\ \dots & \dots \\ 0 & 0 \end{pmatrix}.$$

It is obvious that the rank of this matrix is 1. However, the matrix $\binom{\Sigma}{T}$ has the form

$$\begin{pmatrix} 1 & 0 \\ 0 & 0 \\ \dots & \dots \\ 0 & 0 \\ 0 & 1 \\ \dots & \dots \\ 0 & 0 \end{pmatrix}$$

and its rank is 2.

We also have to consider what happens to the period matrix if the variables $u_1, \dots, u_p$ are acted on by a (nonsingular) linear transformation A:

$$u^* = Au, \tag{II.6.8}$$

where $A = (a_{jk})$, $j = 1, \dots, p$; $k = 1, \dots, p$; $|A| \neq 0$. By the formula (II.3.4) we have

$$\Omega^* = (\omega^{(1)^*}, \dots, \omega^{(r)^*}) = (A\omega^{(1)}, \dots, A\omega^{(r)}) = A\Omega. \tag{II.6.9}$$

Consider, in particular, the case when $r \ge p$ and the rank of the (p, r) matrix Ω is equal to the largest possible value p. By rearranging columns (by a transformation from Ω to ΩM, where M is a special unimodular matrix, each row and each column of which contains only one nonzero element equal to 1) one can reduce Ω to a form in which the determinant of the square matrix B formed by the first p columns is different from zero:

$$\Omega = (B, B'). \tag{II.6.10}$$

Here B is a (p, p) matrix, $|B| \neq 0$ and B' is a $(p, r-p)$ matrix. Now let C be any nonsingular (p, p) matrix. We claim that it is possible to make a transformation of variables $u^* = Au$ such that the period matrix has the form

$$\Omega^* = (C, C'), \tag{II.6.11}$$

where C is the given matrix and C' is some $(p, r-p)$ matrix. In order to accomplish this it is enough to take the matrix CB^{-1} as A, where B^{-1} is the inverse of B. But then, because of formula (II.6.9), we get

$$\Omega^* = CB^{-1}\Omega = CB^{-1}(B, B') = (C, CB^{-1}B').$$

It remains only to set $CB^{-1}B' = C'$. In some cases it is convenient, for example, to take $2\pi iE$ as C, where E is the unit (p, p) matrix. We state the result that we have obtained.

THEOREM II.6.1. *Let $f(u)$ be a meromorphic function of p variables that admits p independent complex periods $\omega^{(1)}, \dots, \omega^{(p)}$, and let C be any nonsingular (p, p) matrix. Then by a linear change of variables $u^* = Au$ we can transform $f(u)$ to the form $f^*(u^*)$ so that each column of the matrix C is a period of the transformed function. In particular, we can obtain the conditions*

$$f^*(u_1^*, \dots, u_j^* + 2\pi i, \dots, u_p^*) = f^*(u_1^*, \dots, u_j^*, \dots, u_p^*), \qquad j = 1, \dots, p.$$

§7. Generalized Fourier series of a periodic entire function

Let an entire nondegenerate function $f(u)$ admit p *complex independent* periods $\omega^{(1)}, \dots, \omega^{(p)}$. Here we do not make any a priori assumptions about the fundamental system of periods. It is clear only that it must contain no less than p periods and that the rank of the period matrix is also equal to p. Because of Theorem II.6.1 we can transform $f^*(u^*)$ (by means of an appropriate substitution) to a form such that the conditions

$$f^*(u_1^*, \dots, u_j^* + c_j, \dots, u_p^*) = f^*(u_1^*, \dots, u_j^*, \dots, u_p^*), \qquad j = 1, \dots, p, \tag{II.7.1}$$

hold, where $c_1, \dots, c_p$ are any complex numbers, none of which is equal to 0 (the matrix C in this case is diagonal and c_j are its diagonal elements). We assume that such a substitution has already been made, and do not write stars in the notation for the function and the variables. Thus, we assume that all coordinates of a period $\omega^{(j)}$ are equal to zero except the jth coordinate which is equal to $c_j \neq 0$ $(j = 1, \ldots, p)$. Make a new (nonlinear) change of variables by setting

$$u_j = \frac{c_j}{2\pi i} \operatorname{Ln} \nu_j, \quad \text{i.e.,} \quad v_j = \exp \frac{2\pi i u_j}{c_j}, \; j = 1, \ldots, p. \tag{II.7.2}$$

Then an infinite set of distinct values u_j differing by integer multiples of c_j, corresponds to each value v_j, $v_j \neq 0$. Because of (II.7.1) the function

$$f(u) = f\left(\frac{c_1}{2\pi i} \operatorname{Ln} v_1, \ldots, \frac{c_p}{2\pi i} \operatorname{Ln} v_p\right) = \varphi(v)$$

is single valued and analytic over the p-dimensional space $\mathbf{C}^p$ of complex variables $v_1, \ldots, v_p$, excluding the origin (where the function is not defined). However, for a function of this type there is a *generalized Laurent series expansion* which is derived in the same way as the Laurent series for a function of one variable, except that instead of a single integral of Cauchy type we use a multiple (p-tuple) integral. As a result, we obtain

$$\varphi(v) = \sum_{(\gamma_1, \ldots, \gamma_p)} d_{\gamma_1 \cdots \gamma_p} v_1^{\gamma_1} \cdots v_p^{\gamma_p} = \sum_{(\gamma)} d_\gamma v^\gamma, \tag{II.7.3}$$

where the summation is taken over all possible ordered sets (γ), each of which consists of p integers. The coefficients d_γ are expressed in terms of the sum of the series by formulas of the form

$$d_\gamma = \frac{1}{2\pi i} \int_{|v_1| = \rho_1} dv_1 \quad \text{therefore} \quad \frac{1}{2\pi i} \int_{|v_p| = \rho_p} \frac{\varphi(v)\, dv_p}{v_1^{\gamma_1 + 1} \cdots v_p^{\gamma_p + 1}}, \tag{II.7.4}$$
$$\rho_j > 0, \qquad j = 1, \cdots, p.$$

Therefore two series (II.7.3) with the same sums must have pairwise equal corresponding coefficients. Finally, the series (II.7.3) converges absolutely in $\mathbf{C}^p \backslash \{0\}$ and uniformly in each region of the form $0 < \varepsilon \leq |v_j| \leq 1/\varepsilon$ for an arbitrary $\varepsilon > 0$. Returning to the variables u_j, we obtain *the generalized Fourier series expansion of a periodic entire function* $f(u)$:

$$f(u) = \sum_{(\gamma)} d_\gamma \exp\left[2\pi i \left(\gamma_1 \frac{u_1}{c_1} + \cdots + \gamma_p \frac{u_p}{c_p}\right)\right].$$

Here the exponent under the summation sign can be represented in the matrix form

$$2\pi i(\gamma_i, \ldots, \gamma_p) \begin{pmatrix} \frac{1}{c_1} & \cdots & 0 \\ \cdots & \cdots & \cdots \\ 0 & \cdots & \frac{1}{c_p} \end{pmatrix} \begin{pmatrix} u_1 \\ \vdots \\ u_p \end{pmatrix} = 2\pi i \tilde{\gamma} C^{-1} u,$$

where $\tilde{\gamma}$ means that the column γ is changed into a row (transposed) and C^{-1} is the matrix inverse to C. Finally we get

$$f(u) = \sum_{(\gamma)} d_\gamma \exp(2\pi i \tilde{\gamma} C^{-1} u). \tag{II.7.5}$$

However, the properties of the generalized Laurent series (II.7.3) translate into properties of the generalized Fourier series in the following way: (a) *the series* (II.7.5) *converges absolutely on* $\mathbf{C}^p$ *and uniformly in each p-disk* $|u_j| \le R$ $(j = 1, \dots, p)$; (b) *from the equality of the sums of two series of the form* (II.7.5) (*with the same matrix* C) *it follows that their corresponding coefficients are pairwise equal.*

The expansion (II.7.5) takes an especially simple form in the case when $C = 2\pi i E_p$, where E_p is a (p, p) unit matrix. Then we obtain

$$f(u) = \sum_{(\gamma)} d_\gamma \exp(\tilde{\gamma} u). \tag{II.7.6}$$

With the help of the expansion we obtained it is possible to show, for example, that *a nondegenerate entire function which has p complex independent periods cannot have any periods independent of them.* Recall that the term independence without additional qualifiers means *integral independence.*

Assume that under the conditions of the theorem there is one more period ω such that from the condition

$$m_1 \omega^{(1)} + \dots + m_p \omega^{(p)} + m\omega = 0,$$

where $m_1, \dots, m_p$ and m are integers, it follows that $m_1 = \dots = m_p = m = 0$. Assuming that from the very beginning the columns of the diagonal matrix $C = 2\pi i E_p$ are periods $\omega^{(j)}$, we substitute $u + \omega$ for u in (II.7.6). We obtain

$$f(u + \omega) = \sum_{(\gamma)} d_\gamma \exp[\tilde{\gamma}(u + \omega)] = f(u) = \sum_{(\gamma)} d_\gamma \exp(\tilde{\gamma} u),$$

from which, in view of the uniqueness of the generalized Fourier series expansion, we get

$$d_\gamma \exp(\tilde{\gamma}\omega) = d_\gamma \quad \text{or} \quad d_\gamma [1 - \exp(\tilde{\gamma}\omega)] = 0 \tag{II.7.7}$$

for all multi-indices γ. We retain from the equalities (II.7.7) only those for which $d_\gamma \ne 0$. We get, for the corresponding multi-indices γ,

$$\exp(\tilde{\gamma}\omega) = 1 \quad \text{or} \quad \gamma_1 \omega_1 + \dots + \gamma_p \omega_p = 2\pi i n_\gamma, \tag{II.7.8}$$

where n_γ are integers and $\omega_1, \dots, \omega_p$ are the complex coordinates of ω.

Let us assume that the set of indices γ for which equalities of the form (II.7.8) hold is rich enough so that we are able to choose p indices $\gamma^{(1)}, \dots, \gamma^{(p)}$ for which the matrix $(\gamma_k^{(j)})$ $(j = 1, \dots, p;\ k = 1, \dots, p)$ of their coordinates is nonsingular. Then from the system of equations

$$\gamma_1^{(j)} \omega_1 + \dots + \gamma_p^{(j)} \omega_p = 2\pi i n_j \qquad (j = 1, \dots, p)$$

we derive $\omega_j = 2\pi i \mu_j / \nu_j$ $(j = 1, \dots, p)$, where μ_j and ν_j are integers; by reducing μ_j/ν_j to a common denominator we have

$$\omega_j = 2\pi i \frac{m_j}{n} \qquad (j = 1, \dots, p).$$

But this means that

$$m_1\omega^{(1)} + \cdots + m_p\omega^{(p)} - n\omega = 0,$$

where $n \neq 0$, i.e., the periods $\omega^{(1)}, \ldots, \omega^{(p)}$ and ω turn out to be *dependent* (recall that, according to our condition $C = 2\pi i E_p$, the period $\omega^{(j)}$ is a column all of whose elements are 0 except the element $2\pi i$ in the jth place).

It remains to show that among the vectors $\gamma = (\gamma_1 \cdots \gamma_p)^{\mathrm{T}}$ with integer coordinates for which the equalities (II.7.8) hold one can actually find p independent vectors. Suppose that this is false and let the maximal number q of independent vectors be less than p. Denote these vectors by $\gamma^{(1)}, \ldots, \gamma^{(q)}$. Then for any vector γ for which the coefficient d_γ in the series (II.7.6) is different from zero, we have

$$m_1\gamma^{(1)} + \cdots + m_q\gamma^{(q)} - n\gamma = 0, \tag{II.7.9}$$

where $m_1, \ldots, m_q$ and n are integers and $n \neq 0$. The corresponding term of the series (II.7.6) can be written in the form

$$d_\gamma \exp(\tilde{\gamma}u) = d_\gamma \exp\left(\sum_{j=1}^{q} \frac{m_j}{n}\tilde{\gamma}^{(j)}u\right).$$

Set $\tilde{\gamma}^{(j)}u = v_j$, $j = 1, \ldots, q$. Then the expansion (II.7.6) takes the form

$$f(u) = \sum d_\gamma \exp\left(\frac{1}{n}\sum_{j=1}^{q} m_j v_j\right) = f^*(v).$$

Here n, m_j are integers depending on γ and determined by the conditions (II.7.9). We see that $f(u)$ becomes a function of $q < p$ variables and therefore is a *degenerate* function if we consider it as a function of the variables v_j. But this contradicts the assumption.

The theorem thus proved is a particular case of another, which we will, however, not use in this book. *A nondegenerate entire function $f(u)$ of p variables cannot have more than p independent periods.*

§8. Construction of an entire function of one variable from its difference

If τ is a p-dimensional vector and $f(u)$ is an entire function of p variables, then

$$f(u+\tau) - f(u) = \Delta_\tau f(u) \tag{II.8.1}$$

is an entire function of u and it is natural to call (II.8.1) a (*first*) *difference of* $f(u)$ with respect to the "step" τ. One of the problems which we will encounter in later chapters is the *construction of an entire function* $f(u)$ from a system of its differences $\Delta_{\tau^{(j)}} f(u) = h_j(u)$, $j = 1, \ldots, p$, where $\tau^{(j)}$, $j = 1, \ldots, p$, are *complex independent vectors*. It is obvious, that this problem does not have a unique solution. If $f_0(u)$ is one of its solutions then the general solution $f(u)$ has the form $f(u) = f_0(u) + P(u)$, where $P(u)$ is any entire function which has p complex independent periods $\tau^{(j)}$ $(j = 1, \ldots, p)$. In this section we consider only the simplest special case of this problem when $p = 1$ and when therefore τ is some nonzero complex number. Thus our immediate goal is *the construction of a particular solution* $f_0(u)$ *of a difference equation*

$$f(u+\tau) - f(u) = h(u), \tag{II.8.2}$$

where $h(u)$ is a given entire function of one variable u.

Since $h(u) = \sum_0^\infty h_n u^n$, it is natural (using the linearity of the problem) to begin by considering an equation of the special form

$$f(u+\tau) - f(u) = u^n, \tag{II.8.3}$$

and then to construct a solution of the equation (II.8.2) from the solutions of equations of the form (II.8.3) when $n = 0, 1, 2, \ldots$. In solving this problem we follow A. Hurwitz (with some simplifications in the computation).

To begin with, by substituting $u\tau$ for u in (II.8.2) we reduce this equation to the simple form

$$f(u+1) - f(u) = h(u) \tag{II.8.2$'$}$$

(here $f(u)$ and $h(u)$ denote the functions which would originally have been written as $f(\tau u)$ and $h(\tau u)$). Equation (II.8.3) then becomes

$$f(u+1) - f(u) = u^n. \tag{II.8.3$'$}$$

It is easy to find a particular solution of this equation in the form of a polynomial of degree $n+1$. This is the *Bernoulli polynomial* $B_{n+1}(u)$ (determined up to a constant term). However, we are not able to construct a solution of the equation (II.8.2$'$) by using the Bernoulli polynomials since in general the series obtained from $h(u) = \sum_0^\infty h_n u^n$ by substituting $B_{n+1}(u)$ for u^n is not convergent. However, the series $\sum_0^\infty h_n B_{n+1}(u)$ formally satisfies (II.8.2). Consequently A. Hurwitz suggested starting with a particular solution of (II.8.3) which is the sum of a Bernoulli polynomial and some finite combination of exponential functions. The form of Hurwitz's solution is suggested by the generating function of the Bernoulli polynomials. It is easy to verify directly that the generating function has the form

$$\Phi(u, z) = \frac{e^{uz} - 1}{e^z - 1} = \sum_0^\infty \frac{B_{n+1}(u)}{n!} z^n \qquad (|z| < 2\pi).$$

If we now notice that $\Phi(u+1, z) - \Phi(u, z) = e^{zu}$ and compare the coefficients on the two sides of the preceding relation we will find that $B_{n+1}(u+1) - B_{n+1}(u) = u^n$, i.e., that $B_{n+1}(u)$ is a particular solution of the equation (II.8.3′).

However, we do not need to go into details here. Set (following Hurwitz)

$$\psi_n(u) = \frac{n!}{2\pi i} \int_{\Gamma_n \,:\, |z|=(2n+1)\pi} \frac{e^{uz} - 1}{e^{z-1}} \frac{dz}{z^{n+1}}. \tag{II.8.4}$$

It is obvious that this is an entire function of u and that, using the residue theorem, it can easily be expressed as a finite linear combination of $B_{n+1}(u)$ and exponential functions with exponents of the form $2k\pi iu$ $(k = 0, \pm 1, \dots, \pm n)$. However, the calculations have no bearing on what follows and therefore we omit them.

By definition,

$$\psi_n(u+1) - \psi(u) = \frac{n!}{2\pi i} \int_{\Gamma_n} e^{uz} \frac{dz}{z^{n+1}} = u^n, \tag{II.8.5}$$

i.e., $\psi_n(u)$ is a particular solution of the equation (II.8.3′). We now show that a particular solution of (II.8.2′) can be obtained by substituting $\psi_n(u)$ for u^n in the expansion $h(u) = \sum_0^\infty h_n u^n$. The Taylor coefficients of any entire function $h(u)$ have the property that the series $\sum_0^\infty |h_n| \rho^n$ converges for each $\rho > 0$. Hence to prove that the series $\sum h_n \psi_n(u)$ is absolutely convergent for each u, uniformly in any disk $|u| \le R$, $R > 0$, it is enough to show that

$$|\psi_n(u)| \le C(R)\rho^n(R) \quad \text{when} \quad |u| \le R.$$

By (II.8.4), we have

$$|\psi_n(u)| \le n! \frac{\max_{\Gamma_n} |e^{uz} - 1|}{\min_{\Gamma_n} |e^z - 1|} \frac{(2n+1)\pi}{[(2n+1)\pi]^{n+1}},$$

or

$$|\psi_n(u)| \le \frac{n!}{(2n+1)^n \pi^n} \frac{M_n}{m_n}, \tag{II.8.6}$$

where

$$M_n = \max_{\Gamma_n} |e^{uz} - 1|, \qquad m_n = \min_{\Gamma_n} |e^z - 1|.$$

But

$$\frac{n!}{(2n+1)^n} = \frac{1 \cdot 2 \cdots n}{(2n+1) \cdots (2n+1)} < \frac{1}{2^n}, \quad M_n < e^{R(2n+1)\pi} + 1 < 2e^{R\pi}(e^{2R\pi})^n, \tag{II.8.7}$$

if $|u| < R$.

To estimate m_n we notice that, at the points Γ_n which belong to the neighborhoods $|z-\zeta| < 1/2$, where $\zeta = \pm(2n+1)\pi i$, we have

$$\begin{aligned}|e^z - 1| &\geq |e^\zeta - 1| - |e^z - e^\zeta| \\ &= 2 - |e^{z-\zeta} - 1| \qquad (e^\zeta = -1).\end{aligned}$$

Since

$$\begin{aligned}|e^{z-\zeta} - 1| &= \left|(z-\zeta) + \frac{(z-\zeta)^2}{2!} + \cdots\right| \\ &\leq |z-\zeta|\left(1 + \frac{1}{2} + \frac{1}{2^2} + \cdots\right) = 2|z-\zeta|,\end{aligned}$$

we have

$$|e^z - 1| > 1, \quad \text{if } |z-\zeta| < \frac{1}{2},\ z \in \Gamma_n. \tag{II.8.8}$$

The circles $|z-\zeta| = 1/2$ intersect Γ_n at the points with abscissas $\pm\xi_n$. It is obvious (Figure 9) that $|\xi_n|$ increases with n and tends to the limit $1/2$. Hence for the points $z \in \Gamma_n$ outside the indicated neighborhoods we have

$$\begin{aligned}|e^z - 1| &\geq ||e^z| - 1| = |e^x - 1| > 1 - e^{-|\xi_n|} \\ &\geq 1 - e^{-|\xi_1|} = \delta > 0.\end{aligned}$$

Thus

$$m_n \geq \delta > 0. \tag{II.8.9}$$

From the inequalities (II.8.6), (II.8.7), and (II.8.9) it now follows that

$$|\psi_n(u)| \leq \frac{2e^{R\pi}}{\delta}\left(\frac{e^{2R\pi}}{2\pi}\right)^n = C(R)\rho^n(R) \quad \text{when } |u| \leq R,$$

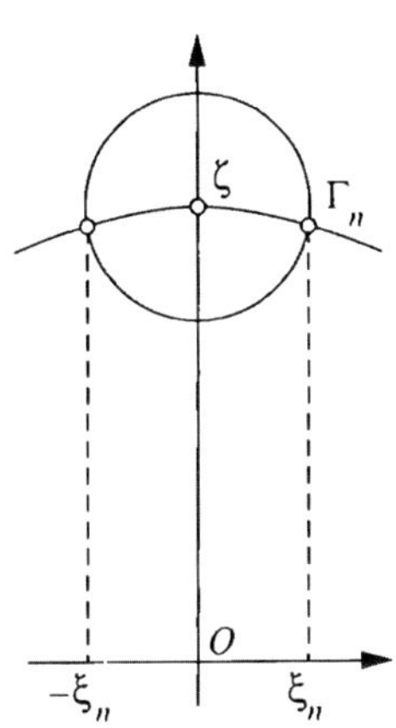

FIGURE 9

where $\rho(R) = e^{2\pi R}/2\pi$. Hence we have proved that the series $\sum_0^\infty h_n \psi_n(n)$ represents an entire function $f_0(u)$ satisfying (II.8.2′), namely $f_0(u) = \sum_0^\infty h_n \psi_n(u)$. Therefore the entire function

$$f_0(u) = \sum_0^\infty h_n \tau^n \psi_n\left(\frac{u}{\tau}\right)$$

satisfies the equation (II.8.2). To obtain the general solution of (II.8.2) (in terms of entire functions) it suffices to add an arbitrary entire function with period τ.

CHAPTER III

Riemann Matrices. Jacobian (Intermediate) Functions

§1. Riemann matrices. Elementary conditions

This chapter and the next are devoted to the Abelian functions discussed in section I.9.

An *Abelian function* is a (nondegenerate) meromorphic function of p complex variables: $u_1, \ldots, u_p$ with $2p$ independent periods $\omega^{(1)}, \ldots, \omega^{(2p)}$.

In section II.4 we gave an example of a function satisfying these conditions for every $p \geq 1$. Therefore Abelian functions exist. However, we call the reader's attention to the fact that this example contributes little to the understanding of Abelian functions. Indeed, the function previously mentioned is simply a linear combination of elliptic functions which are doubly periodic in each of p variables separately. As a consequence, it has none of the singular points of indeterminacy which meromorphic functions have for $p > 1$ (compare section II.2). Moreover, it has two periods for each variable separately (when the rest are held fixed), whereas in the general case an Abelian function need not have such properties (see the end of this section).

Turning to the general study of Abelian functions, we place the most emphasis on the study of the properties of the matrix of the system of fundamental periods of an Abelian function.

A matrix

$$\Omega(\omega^{(1)}, \ldots, \omega^{(2p)}) = \begin{pmatrix} \omega_1^{(1)} & \cdots & \omega_1^{(2p)} \\ \cdots & \cdots & \cdots \\ \omega_p^{(1)} & \cdots & \omega_p^{(2p)} \end{pmatrix} \qquad \text{(III.1.1)}$$

is called a *Riemann matrix* if there exists an Abelian function for which the columns of the matrix form a fundamental system of periods.

A very special example of a Riemann matrix was given in section II.2 in connection with the example just mentioned. This matrix has the special form

$$\Omega = \begin{pmatrix} 1 & 2\tau & \cdots & 0 & 0 \\ 0 & 0 & \cdots & 0 & 0 \\ \cdots & \cdots & \cdots & \cdots & \cdots \\ 0 & 0 & \cdots & 1 & 2\tau \end{pmatrix},$$

where $\operatorname{Im}\tau > 0$. It is not at all trivial to find conditions which are simultaneously necessary and sufficient for a $(p, 2p)$ matrix with complex elements to be a Riemann matrix. We shall obtain such conditions as the end result of the analysis described in this chapter. Starting from these conditions we will be able to construct the corresponding Abelian functions relatively easily (using, however, an appropriate generalization of the analytic apparatus developed by Jacobi for the case $p = 1$).

It is natural to establish an *equivalence relation* for Riemann matrices by considering two matrices Ω_1 and Ω_2 of the form (III.1.1) to be in the same class if and only if there is a nonsingular linear transformation A with matrix (a_{jk}), $j = 1, \dots, p$; $k = 1, \dots, p$, where a_{jk} are complex numbers, and there is a unimodular transformation M with matrix $(m_{\alpha\beta})$, $\alpha = 1, \dots, 2p$; $\beta = 1, \dots, 2p$, where $m_{\alpha\beta}$ are real numbers, such that

$$\Omega_2 = A\Omega_1 M. \tag{III.1.2}$$

The reader probably remembers (section II.6) that the transformation M transforms a fundamental system Ω_1 to a system $\Omega_1^* = \Omega_1 M$, and the transformation A transforms the variables $u_1, \dots, u_p$ to new variables $u_1^*, \dots, u_p^*$ by means of the formula $u^* = Au$. Consequently, the period matrix Ω_1^* is transformed into $\Omega_2 = A\Omega_1^*$. It is easy to verify that the relation (III.1.2) just introduced between $(2p, p)$ matrices really has all the properties of an equivalence relation.

Consequently, we can always use a transformation of the form (III.1.2) to reduce a Riemann matrix to simplest form without leaving a given equivalence class. But before doing this we notice the so-called *elementary conditions* which must be satisfied by every Riemann matrix (these conditions turn out to be simple consequences of the necessary and sufficient conditions which will be formulated in the section IV.1). These state that each Riemann matrix (III.1.1) must satisfy the following conditions:

(1) *None of its rows can contain only real or only pure imaginary numbers; in particular, none of its rows can consist only of zeros.*

(2) *None of its columns can consist only of zeros.*

(3) *The rank of the matrix is the same as the number* p *of its rows.*

These follow from the fact that the periods $\omega^{(k)}$, $k = 1, \dots, 2p$, are linearly independent over the reals. Indeed, if we set $\omega_j^{(k)} = \alpha_j^{(k)} + i\beta_j^{(k)}$ where $\alpha_j^{(k)}$ and $\beta_j^{(k)}$ are real numbers, then Ω can be written in the form

$$\Omega = \Omega' + i\Omega'', \tag{III.1.3}$$

where

$$\Omega' = \begin{pmatrix} \alpha_1^{(1)} & \cdots & \alpha_1^{(2p)} \\ \cdots & \cdots & \cdots \\ \alpha_p^{(1)} & \cdots & \alpha_p^{(2p)} \end{pmatrix}, \qquad \Omega'' = \begin{pmatrix} \beta_1^{(1)} & \cdots & \beta_1^{(2p)} \\ \cdots & \cdots & \cdots \\ \beta_p^{(1)} & \cdots & \beta_p^{(2p)} \end{pmatrix}.$$

We form the $(2p, 2p)$ matrix

$$\begin{pmatrix} \Omega' \\ \Omega'' \end{pmatrix}. \tag{III.1.4}$$

Its kth column consists of $2p$ real coordinates of the vector $\omega^{(k)}$, considered as a vector in $\mathbf{R}^{2p}$. Because of the independence of these vectors (which follows (see section II.4) from the independence of the corresponding periods) the matrix (III.1.4) cannot be singular, i.e.,

$$\begin{vmatrix} \Omega' \\ \Omega'' \end{vmatrix} = D \neq 0. \tag{III.1.5}$$

In case one of the conditions (1) or (2) does not hold, one row or one column of the preceding determinant consists only of zeros, contradicting (III.1.5).

In order to show that condition (3) is necessary we also compute the determinant $|F|$ defined by the relation

$$|F| = \begin{vmatrix} \Omega \\ \overline{\Omega} \end{vmatrix} = \begin{vmatrix} \Omega' & +i\Omega'' \\ \Omega' & -i\Omega'' \end{vmatrix}. \tag{III.1.6}$$

This determinant can be represented in the form

$$|F| = \begin{vmatrix} E_p & iE_p \\ E_p & -iE_p \end{vmatrix} \cdot \begin{vmatrix} \Omega' \\ \Omega'' \end{vmatrix},$$

where E_p is the unit (p, p) matrix. And since (if we subtract the first "row" from the second)

$$\begin{vmatrix} E_p & iE_p \\ E_p & -iE_p \end{vmatrix} = \begin{vmatrix} E_p & iE_p \\ 0 & -2iE_p \end{vmatrix} = (-2i)^p,$$

we have

$$|F| = (-2i)^p D \neq 0. \tag{III.1.7}$$

Comparison of this result with (III.1.6) shows that the rank of the matrix Ω is not less than p (otherwise we would have $|F| = 0$). Because this rank cannot be greater than p, property (3) is also proved. Because of property (3) there must be p complex independent columns among the columns of a Riemann matrix. By changing the numbering of the periods (which can be considered as a result of an appropriate unimodular transformation preserving the original equivalence class) we can assume that the determinant formed by the first p columns of the matrix Ω is different from zero.

Setting

$$\begin{pmatrix} \omega_1^{(1)} & \cdots & \omega_1^{(p)} \\ \cdots & \cdots & \cdots \\ \omega_p^{(1)} & \cdots & \omega_p^{(p)} \end{pmatrix} = \Sigma, \qquad \begin{pmatrix} \omega_1^{(p+1)} & \cdots & \omega_1^{(2p)} \\ \cdots & \cdots & \cdots \\ \omega_p^{(p+1)} & \cdots & \omega^{(2p)} \end{pmatrix} = T,$$

we have

$$\Omega = (\Sigma, T), \tag{III.1.8}$$

where $|\Sigma| \neq 0$.

Now let Σ^* be any nonsingular (p, p) matrix. Applying the linear transformation

$$u^* = Au = \Sigma^*\Sigma^{-1}u, \tag{III.1.9}$$

to the independent variables, we reduce Ω to the equivalent form

$$\Omega^* = (\Sigma^*, T^*), \tag{III.1.10}$$

where Σ^* is given in advance and $T^* = \Sigma^*\Sigma^{-1}T$. In particular, taking $\Sigma^* = 2\pi i E_p$ we can assert that *each equivalence class of Riemann matrices must contain a matrix of the form*

$$\Omega = (2\pi i E_p, T). \tag{III.1.11}$$

If we set

$$T = \begin{pmatrix} t_{11} & \cdots & t_{1p} \\ \cdots & \cdots & \cdots \\ t_{p1} & \cdots & t_{pp} \end{pmatrix},$$

then the properties of periodicity of an Abelian function $f(u)$ corresponding to the matrix (III.1.11) can be written in the following form:

$$\begin{aligned} &f(u_1, \dots, u_{j-1}, u_j + 2\pi i, u_{j+1}, \dots, u_p) \\ &\qquad = f(u_1, \dots, u_{j-1}, u_j, u_j, \dots, u_p), \qquad j = 1, \dots, p; \\ &f(u_1 + t_{1k}, \dots, u_p + t_{pk}) = f(u_1, \dots, u_p), \qquad k = 1, \dots, p. \end{aligned} \tag{II.1.12}$$

In other words: *an Abelian function* $f(u)$ (after an appropriate change of variables which corresponds to transition to the matrix of the form (III.1.11)) *is periodic with period* $2\pi i$ *in each of the variables* u_j *separately* (the rest are then fixed). *The remaining* p *periods of this function* $t^{(1)}, \dots, t^{(p)}$, *which are expressed by the columns of the matrix* T, *require a simultaneous change of all* p *variables* u_j *by the corresponding addends* t_{jk} *that are in general nonzero.*

Some properties of the matrix T which ultimately follow from (III.1.5) will already be needed in the following sections. Let us first write the matrix (III.1.11) in the form (III.1.3):

$$\Omega = \Omega' + i\Omega'' = (0, T') + i(2\pi E_p, T''),$$

where the elements of the matrices T' and T'' are real numbers ($T = T' + iT''$). We will show that T' is *a nonsingular matrix,* i.e.,

$$|T'| \neq 0. \tag{III.1.13}$$

This is the first of the properties needed. In fact, the determinant (III.1.5) can be written in the form

$$D = \begin{vmatrix} \Omega' \\ \Omega'' \end{vmatrix} = \begin{vmatrix} 0 & T' \\ 2\pi E_p & T'' \end{vmatrix} = (-2\pi)^p |T'|,$$

and since $D \neq 0$, we get $|T'| \neq 0$.

The next property is more special. We formulate it as a lemma (it will be used in section III.5).

LEMMA. *There exists a positive number* α *such that for the matrix* T *we have*

$$\max_{1\le k\le p}\left|e^{\tilde{\gamma}t^{(k)}}-1\right|\ge\alpha \tag{III.1.14}$$

for any nonzero vector γ *with integer coordinates.*

PROOF. Assume the contrary. Then for any $\varepsilon>0$ and in particular for $\varepsilon<1/2$ there is a vector $\gamma^{(0)}\ne 0$ with integer coordinates $n_1,\dots,n_p$ (it is obvious that $\|\gamma^{(0)}\|\ge 1$), such that

$$\left|e^{\tilde{\gamma}^{(0)}t^{(k)}}-1\right|<\varepsilon,\qquad k=1,\dots,p. \tag{III.1.15}$$

Let $\operatorname{Re}t_{jk}=t'_{jk}$; then $\eta_k=\operatorname{Re}(\tilde{\gamma}^{(0)}t^{(k)})=n_1t'_{1k}+\dots+n_pt'_{pk}$. We have

$$e^{\eta_k}=\left|e^{\tilde{\gamma}^{(0)}t^{(k)}}\right|=\left|1+\left(e^{\tilde{\gamma}^{(0)}t^{(k)}}-1\right)\right|,$$

from which, by (III.1.15), it follows that $1-\varepsilon<e^{\eta_k}<1+\varepsilon$. Hence $\ln(1-\varepsilon)<\eta_k<\ln(1+\varepsilon)$ which implies (if $0<\varepsilon<1/2$) the inequalities

$$|\eta_k|<2\varepsilon,\qquad k=1,\dots,p.$$

Furthermore, $T'\gamma^{(0)}=\eta$ where η is a vector with coordinates $\eta_1,\dots,\eta_p$ (therefore $\|\eta\|<2\varepsilon$). Because of (III.1.13), $\gamma^{(0)}=(T')^{-1}\eta$, from which it follows that

$$1\le\|\gamma^{(0)}\|\le\|(T')^{-1}\|\,\|\eta\|<c_0\varepsilon,\qquad c_0=\text{const}. \tag{III.1.16}$$

When $\varepsilon<1/c_0$ the inequalities (III.1.16) are contradictory; this establishes the lemma.

§2. The first system of difference equations. Conditions of solvability

We will denote by $e^{(h)}$ the p-vector all of whose coordinates are zero except the hth, which is equal to 1. Then we can write, for example,

$$E_p=(e^{(1)},\dots,e^{(p)})\quad\text{or}\quad 2\pi iE_p=(2\pi ie^{(1)},\dots,2\pi ie^{(p)}).$$

If $f(u)$ is an Abelian function whose Riemann matrix Ω has the form

$$\Omega=(2\pi iE_p,T), \tag{III.2.1}$$

then $f(u+2\pi ie^{(h)})=f(u)$, $h=1,\dots,p$.

Let

$$f(u)=\frac{\varphi(u)}{\psi(u)} \tag{III.2.2}$$

be a reduced representation of this function. By definition (see section II.2), $\varphi(u)$ and $\psi(u)$ are entire functions which vanish simultaneously at the points of indeterminacy of the function $f(u)$ and only at those points. If ω is a period of $f(u)$, then

$$f(u+\omega) = \frac{\varphi(u+\omega)}{\psi(u+\omega)} = f(u). \tag{III.2.3}$$

Hence formula (III.2.3) gives an additional representation of $f(u)$ as a quotient of two entire functions $\varphi(u+\omega)$ and $\psi(u+\omega)$, for which ω is not necessarily a period. But, because of the periodicity of $f(u)$, the point $u^{(0)}+\omega$ is a point of indeterminacy at the same time as the point $u^{(0)}$. This is because if we can find a sequence $\{u^{(n)}\}$ in a neighborhood of the point $u^{(0)}$ such that for an arbitrary complex number c we have $\lim_{u^{(n)}\to u^{(0)}} f(u^{(n)}) = c$, then the sequence $\{u^{(n)}+\omega\}$ will play an analogous role in a neighborhood of the point $u^{(0)}+\omega$ and

$$\lim_{u^{(n)}+\omega\to u^{(0)}+\omega} f(u^{(n)}+\omega) = c,$$

since $f(u^{(n)}+\omega) = f(u^{(n)})$. Hence $\varphi(u^{(0)}+\omega) = 0$ and $\psi(u^{(0)}+\omega) = 0$; in other words, the following pairs of equations are valid simultaneously:

$$\varphi(u^{(0)}) = 0, \qquad \psi(u^{(0)}) = 0,$$

and

$$\varphi(u^{(0)}+\omega) = 0, \qquad \psi(u^{(0)}+\omega) = 0.$$

Thus if the representation (III.2.2) is reduced, then (III.2.3) is also reduced, and vice versa.

But $2\pi i e^{(h)}$ $(h = 1, \dots, p)$ are periods of $f(u)$ and hence we get p additional reduced representations of $f(u)$:

$$f(u) = \frac{\varphi(u+2\pi i e^{(h)})}{\psi(u+2\pi i e^{(h)})} \qquad (h = 1, \dots, p). \tag{III.2.4}$$

Comparing (III.2.4) with (III.2.2), we obtain, by section II.2,

$$\varphi(u+2\pi i e^{(h)}) = e^{g_h(u)}\varphi(u), \qquad \psi(u+2\pi i e^{(h)}) = e^{g_h(u)}\psi(u), \\ h = 1, \dots, p, \tag{III.2.5}$$

where $g_h(u)$ are entire functions. Thus *to each representation of the form* (III.2.2) *there correspond* p *entire functions* $g_h(u)$ *determined by the equations* (III.2.5) *up to arbitrary integer multiples of* $2\pi i$.

Note that the functions g_h and g_k are interrelared by their definitions. Indeed, if we set u equal to $u+2\pi i e^{(k)}$ in the first of relation (III.2.5), we obtain

$$\varphi(u+2\pi i e^{(h)}+2\pi i e^{(k)}) = e^{g_h(u+2\pi i e^{(k)})}\varphi(u+2\pi i e^{(k)}) = e^{g_h(u+2\pi i e^{(k)})+g_k(u)}\varphi(u).$$

Now we interchange the indices h and k to get

$$\varphi(u+2\pi i e^{(k)}+2\pi i e^{(h)}) = e^{g_k(u+2\pi i e^{(h)})+g_h(u)}\varphi(u).$$

Comparing these results, we find

$$[g_k(u+2\pi i e^{(h)}) - g_k(u)] - [g_h(u+2\pi i e^{(k)}) - g_h(u)] \\ = 2\pi i n_{hk}, \tag{III.2.6}$$

where n_{nk} $(h = 1, \dots, p;\ k = 1, \dots, p)$ must be integers. From the relations (III.2.6) it follows further that $n_{hk} = -n_{kh}$ and, in particular, $n_{hh} = 0$. Thus the square matrix

$$N = (n_{hk}), \qquad h = 1, \dots, p, \ k = 1, \dots, p, \tag{III.2.7}$$

is skew-symmetric. We will see later that it does not depend on the choice of the representation (III.2.2).

We will try to use the fact that there are many representations of the form (III.2.2) in order to arrive at the simplest possible set of $g_h(u)$ consisting only of *linear functions.* Since any reduced representation is obtained from (III.2.2) by multiplication of the numerator and denominator by $\exp G(u)$ where $G(u)$ is an entire function, everything is reduced to finding an appropriate entire function $G(u)$.

So let

$$\varphi^*(u) = e^{G(u)}\varphi(u), \qquad \psi^*(u) = e^{G(u)}\psi(u). \tag{III.2.8}$$

Then, instead of (III.2.5), we will have the following form for the new representation:

$$\varphi^*(u+2\pi i e^{(h)}) = e^{g_h^*(u)}\varphi^*(u), \qquad \psi^*(u+2\pi i e^{(h)}) = e^{g_h^*(u)}\psi^*(u). \tag{III.2.9}$$

As a consequence, the numerator and denominator of the representation are changed in exactly the same manner, and from now on we will follow only the changes of the numerator. From (III.2.8) and (III.2.5) we derive

$$\begin{aligned} \varphi^*(u+2\pi i e^{(h)}) &= \exp[G(u+2\pi i e^{(h)})]\varphi(u+2\pi i e^{(h)}) \\ &= \exp[G(u+2\pi i e^{(h)}) + g_h(u)]\varphi(u) \\ &= \exp[G(u+2\pi i e^{(h)}) - G(u) + g_h(u)]\varphi^*(u); \end{aligned}$$

comparison with (III.2.9) gives

$$g_h^*(u) = G(u+2\pi i e^{(h)}) - G(u) + g_h(u), \qquad h = 1, \dots, p$$

(here we used the fact that the function $g_h^*(u)$ is determined by the first of the equations (III.2.9) only up to an additive multiple of $2\pi i$). Thus if we pass from the functions $\varphi(u)$ and $\psi(u)$ to the functions (III.2.8), the functions $g_h(u)$ change by the corresponding differences $G(u + 2\pi i e^{(h)}) - G(u)$. In particular, each of the formulas in square brackets on the right-hand side of (III.2.6) changes in the same way by the value of the difference

$$[G(u+2\pi i e^{(h)} + 2\pi i e^{(k)}) - G(u+2\pi i e^{(h)})] - [G(u+2\pi i e^{(k)}) - G(u)],$$

from which it follows that the *integers* n_{hk} *are invariant under a change from* $\varphi(u)$ *and* $\psi(u)$ *to* $\varphi^*(u)$ *and* $\psi^*(u)$.

In order for all $g_h^*(u)$ to be linear, that is,

$$g_h^*(u) = l_h(u) = \sum_{k=1}^{p} \lambda_{hk} u_k , \tag{III.2.10}$$

it is enough to require that the partial differences of the function $G(u)$ that we are looking for must be determined by equations of the form

$$G(u + 2\pi i e^{(h)}) - G(u) = H_h(u), \qquad h = 1, \dots, p, \tag{III.2.11}$$

where $H_h(u)$ are entire functions of the form

$$H_h(u) = l_h(u) - g_h(u), \qquad h = 1, \dots, p. \tag{III.2.12}$$

It is easy to see that for the solutions of the system (III.2.11) to exist, certain *conditions of solvability* must be satisfied. In fact, from (III.2.11) we derive

$$G(u + 2\pi i e^{(h)} + 2\pi i e^{(k)}) - G(u + 2\pi i e^{(k)}) = H_h(u + 2\pi i e^{(k)}),$$

and on the other hand

$$G(u + 2\pi i e^{(k)}) - G(u) = H_k(u).$$

Adding the corresponding sides of the equations, we get

$$G(u + 2\pi i e^{(h)} + 2\pi i e^{(k)}) - G(u) = H_h(u + 2\pi i e^{(k)}) + H_k(u).$$

Let us interchange h and k. Since the left-hand side does not change, it is necessary for the conditions

$$H_h(u + 2\pi i e^{(k)}) + H_k(u) = H_k(u + 2\pi i e^{(h)}) + H_h(u)$$

to hold or

$$H_h(u + 2\pi i e^{(k)}) - H_h(u) = H_k(u + 2\pi i e^{(h)}) - H_k(u). \tag{III.2.13}$$

In other words, *the partial differences of the entire functions* $H_h(u)$

$$H_h(u + 2\pi i e^{(k)}) - H_h(u)$$

must not change when one interchanges the indices h *and* k.

These, then, are the conditions that we need for solvability of the system (III.2.11). We verify that they hold if we introduce the necessary changes in the definition (to make it more precise) of the linear functions (III.2.10). Indeed, from (III.2.12) it follows in view of (III.2.10) that

$$\begin{aligned} H_h(u + 2\pi i e^{(k)}) - H_h(u) &= [l_h(u + 2\pi i e^{(k)}) - l_h(u)] \\ &\quad - [g_h(u + 2\pi i e^{(k)}) - g_h(u)] \\ &= 2\pi i \lambda_{hk} - [g_h(u + 2\pi i e^{(k)}) - g_h(u)]. \end{aligned} \tag{III.2.14}$$

We interchange the roles of the indices h and k here and get

$$H_k(u + 2\pi i e^{(h)}) - H_k(u) = 2\pi i \lambda_{kh} - [g_k(u + 2\pi i e^{(h)}) - g_k(u)]. \tag{III.2.15}$$

The equality of the left-hand sides of (III.2.14) and (III.2.15) therefore assumes the truth of the relations

$$2\pi i(\lambda_{kh} - \lambda_{hk}) - \{[g_k(u + 2\pi i e^{(h)}) - g_k(u)] \\ - g_h(u + 2\pi i e^{(k)}) - g_h(u)]\} = 0,$$

or, in view of (III.2.6),

$$\lambda_{kh} - \lambda_{hk} = n_{hk}, \qquad h = 1, \dots, p; \; k = 1, \dots, p. \tag{III.2.16}$$

The easiest way to satisfy these equalities is by setting

$$\lambda_{hk} = -\frac{1}{2} n_{hk}, \qquad h = 1, \dots, p; \; k = 1, \dots, p. \tag{III.2.17}$$

We will make this choice, and thus complete the definition of the linear functions (III.2.10). We have

$$g_h^*(u) = l_h(u) = -\frac{1}{2}\sum_{k=1}^{p} n_{hk} u_k . \tag{III.2.10$'$}$$

Thus, *the coefficients of the function* $2l_h(u)$ *are indeed real numbers; they are elements of a skew-symmetric matrix* N *which depends only on the Abelian function* $f(u)$ *but not on its particular representation as a quotient.*

Since the conditions of solvability (III.2.13) are now satisfied we can turn to the search for an entire function $G(u)$ which is a solution of the first system of difference equations (III.2.11).

§3. Construction of the solution of the first system

It was shown in section II.8 that an equation of the form

$$G(u + \tau) - G(u) = H(u) = \sum_0^\infty c_n u^n,$$

where u is a complex variable $(p = 1)$, and $H(u)$ is an entire function, has as a particular solution a function

$$G_0(u) = \sum_0^\infty c_n \tau^n \psi_n\left(\frac{u}{\tau}\right);$$

entire analytic functions $\psi_n(u/\tau)$ which do not depend on $H(u)$ were defined in the same section. We apply this result to an equation of the form

$$G(u_1, \dots, u_h + 2\pi i, \dots, u_p) - G(u_1, \dots, u_h, \dots, u_p) \\ = F_h(u_1, \dots, u_h, \dots, u_p), \tag{III.3.1}$$

where $F_h(u_1, \dots, u_h, \dots, u_p)$ is an entire function. Here we will assume that the values of all variables except u_h are arbitrarily fixed. We note that a decomposition of F_h in powers of u_h has the form

$$F_h(u) = \sum_0^\infty c_n u_h^n, \tag{III.3.2}$$

where

$$c_n = c_n(u_1, \dots, u_{h-1}, u_{h+1}, \dots, u_p) = \frac{1}{n!}\left(\frac{\partial^n F_h}{\partial u_h^n}\right)_{u_h=0}. \tag{III.3.3}$$

Moreover, the series (III.3.2) converges absolutely and uniformly in any p-disk $|u_j| \le R$, $j = 1, \dots, p$. Applying the result mentioned above to the equation (III.3.1) we get a particular solution of the equation (III.3.1) of the form

$$G_0(u) = \sum c_n(2\pi i)^n \psi_n\left(\frac{u_h}{2\pi i}\right). \tag{III.3.4}$$

We call a solution of this type *canonical*. Because of the estimate for $|\psi_n(z)|$ established in section II.8, the latter series also converges absolutely and uniformly in any p-disk $|u_j| \le R$, $j = 1, \dots, p$, and therefore the series is an entire analytic function of p complex variables u_j, $j = 1, \dots, p$.

Assume now that the function $F_h(u_1, \dots, u_p)$ has period $2\pi i$ in each of the variables $u_1, \dots, u_{h-1}$ when $1 \le h \le p$, i.e.,

$$\begin{aligned} &F_h(u_1 + 2\pi i, u_2, \dots, u_p) \\ &\quad = F_h(u_1, u_2, \dots, u_p), \dots, F_h(u_1, \dots, u_{h-1} + 2\pi i, \dots, u_p) \\ &\quad = F_h(u_1, \dots, u_p). \end{aligned}$$

Then we can assert that the canonical solution (III.3.4) has the same properties of periodicity with respect to these variables. Indeed, as the expansion (III.3.2) we have

$$\begin{aligned} F_h(u + 2\pi i e^{(j)}) &= \sum_{n=0}^{\infty} c_n(u_1, \dots, u_j + 2\pi i, \dots, u_{h-1}, u_{h+1}, \dots, u_p)u_h^n \\ &= F_h(u) = \sum_{n=0}^{\infty} c_n(u_1, \dots, u_j, \dots, u_{h-1}, u_{h+1}, \dots, u_p)u_h^n \\ &\qquad (j = 1, \dots, h-1). \end{aligned}$$

Since the decomposition into a power series is unique it follows that

$$c_n(u_1, \dots, u_j + 2\pi i, \dots, u_p) = c_n(u_1, \dots, u_j, \dots, u_p) \qquad (j = 1, \dots, h-1),$$

i.e., the functions c_n have period $2\pi i$ in each of the variables $u_1, \dots, u_{h-1}$. Hence the function $G_0(u)$ (the sum of the series (III.3.4)) also has these properties.

We now solve the equations of the system (III.2.11) successively, beginning with the first. Let $G_1(u)$ be a canonical solution of the equation

$$G(u_1 + 2\pi i, u_2, \dots, u_p) - G(u_1, u_2, \dots, u_p) = H_1(u_1, \dots, u_p). \tag{III.3.5}$$

Then the general solution of this equation in terms of entire functions has the form

$$G(u) = G_1(u) + P(u), \tag{III.3.6}$$

where $P(u)$ is any entire function of period $2\pi i$ in the variable u_1:

$$P(u_1 + 2\pi i, u_2, \dots, u_p) = P(u_1, u_2, \dots, u_p). \tag{III.3.7}$$

Let us substitute (III.3.6) into the second equation of the system (III.2.11). We obtain the following equation, which determines an entire function $P(u)$:

$$P(u_1, u_2 + 2\pi i, \dots, u_p) - P(u_1, u_2, \dots, u_p) = F_2(u_1, u_2, \dots, u_p), \tag{III.3.8}$$

where

$$F_2(u) = H_2(u) - [G_1(u + 2\pi i e^{(2)}) - G_1(u)].$$

This function has period $2\pi i$ in u_1. In fact because G_1 satisfies (III.3.5) we get

$$\begin{aligned} F_2(u_1 + 2\pi i, u_2, \dots, u_p) &- F_2(u_1, \dots, u_p) \\ &= H_2(u + 2\pi i e^{(1)}) - H_2(u) \\ &\quad - \{[G_1(u + 2\pi i e^{(1)} + 2\pi i e^{(2)}) - G_1(u + 2\pi i e^{(2)})] \\ &\qquad - [G_1(u + 2\pi i e^{(1)}) - G_1(u)]\} \\ &= H_2(u + 2\pi i e^{(1)}) - H_2(u) - [H_1(u + 2\pi i e^{(2)}) - H_1(u)]. \end{aligned}$$

But the last expression is equal to zero because of the solvability conditions (III.2.13). Hence $G_2(u)$ (a canonical solution of the equation (III.3.8)) has period $2\pi i$ for the variable u_1 just as $F_2(u)$ does. Set

$$G(u) = G_1(u) + G_2(u) + P(u), \tag{III.3.9}$$

where $P(u)$ is an entire function of period $2\pi i$ in each of the variables u_1 and u_2. Regardless of which function it is, formula (III.3.9) will give a solution of the first two equations of the system (III.2.11). Therefore we need to choose $P(u)$ so that (III.3.9) also satisfies the third equation of the system.

In what follows, we use induction.

Assume that entire functions $G_n(u)$, $n = 1, \dots, h$, $1 \le h < p$, *have been constructed so that*

(a) $G_n(u+2\pi i e^{(n)}) - G_n(u) = F_n(u) = H_n(u) - \sum_{j=1}^{n-1}[G_j(u+2\pi i e^{(n)}) - G_j(u)]$, i.e., $\sum_{j=1}^{n}[G_j(u + 2\pi i e^{(n)}) - G_j(u)] = H_n(u)$;

(b) $G_j(u + 2\pi i e^{(n)}) - G_j(u) = 0$ provided $j > n$. *Then* $G^{(h)}(u) = \sum_{j=1}^{h} G_j(u)$ *satisfies each of the equations*

$$G^{(h)}(u + 2\pi i e^{(n)}) - G^{(h)} = H_n(u), \qquad n = 1, \dots, h.$$

Indeed,

$$G^{(h)}(u + 2\pi i e^{(n)}) - G^{(h)}(u) = \sum_{j=1}^{h} [G_j(u + 2\pi i e^{(n)}) - G_j(u)].$$

But because of (b) the terms of this sum are equal to zero when $j > n$. The sum

$$\sum_{j=1}^{n} [G_j(u + 2\pi i e^{(n)}) - G_j(u)] = H_n(u)$$

does not cancel, by condition (a).

We seek a solution of the equation

$$G(u + 2\pi i e^{(h+1)}) - G(u) = H_{h+1}(u) \tag{III.3.10}$$

in the form

$$G(u) = \sum_{j=1}^{h} G_j(u) + P(u), \tag{III.3.11}$$

where $P(u)$ has periods $2\pi i e^{(1)}, \dots, 2\pi i e^{(h)}$. It is obvious that for any choice of such an entire function $P(u)$ the function $G(u)$ defined by the last formula satisfies the first h equations of the system (III.2.11). For the $(h+1)$st equation to be satisfied it is necessary for $P(u)$ to satisfy the equation obtained as a result of substituting (III.3.11) in (III.3.10):

$$P(u+2\pi i e^{(h+1)}) - P(u) = F_{h+1}(u) = H_{h+1}(u) - \sum_{j=1}^{h} [G_j(u+2\pi i e^{(h+1)}) - G_j(u)]. \tag{III.3.12}$$

We will show that the entire function $F_{h+1}(u)$ has periods $2\pi i e^{(1)}, \dots, 2\pi i e^{(h)}$. Indeed,

$$\begin{aligned} F_{h+1}(u + 2\pi i e^{(n)}) - F_{h+1}(u) &= H_{h+1}(u + 2\pi i e^{(n)}) - H_{h+1}(u) \\ &- \sum_{j=1}^{h} \{[G_j(u + 2\pi i e^{(h+1)} + 2\pi i e^{(n)}) - G_j(u + 2\pi i e^{(h+1)})] \\ &- [G_j(u + 2\pi i e^{(n)}) - G_j(u)]\} \qquad (1 \le n \le h). \end{aligned}$$

If $j > n$, the corresponding summands in the last sum become zero (by the condition (b)). Hence the summation can be taken from $j = 1$ to $j = n$. We rewrite the right-hand side in the form

$$\begin{aligned} &H_{h+1}(u + 2\pi i e^{(n)}) - H_{h+1}(u) \\ &\quad - \sum_{j=1}^{n} [G_j(u + 2\pi i e^{(h+1)} + 2\pi i e^{(n)}) - G(u + 2\pi i e^{(h+1)})] \\ &\quad + \sum_{j=1}^{n} [G_j(u + 2\pi i e^{(n)}) - G_j(u)]. \end{aligned}$$

But this expression (because of the condition (a) applied to each sum) is equal to

$$[H_{h+1}(u + 2\pi i e^{(n)}) - H_{h+1}(u)] - [H_n(u + 2\pi i e^{(h+1)}) - H_n(u)].$$

Hence, as a consequence of the condition of interchangeability, (III.2.13), it is equal to zero.

Thus $F_{h+1}(u)$ has periods $2\pi i e^{(j)}$, $j = 1, \dots, h$. Hence a canonical solution $P(u) = G_{h+1}(u)$ of the equation (III.3.12) also has the same periods. Adding the functions $G_{h+1}(u)$ and $F_{h+1}(u)$ to the list of functions $G_j(u)$ and $F_j(u)$ $(j = 1, \dots, h)$ found earlier, we can now assert that the conditions (a) and (b) are satisfied for the system extended in this way (h is replaced by $h+1$). Hence, after an appropriate number of steps, we reach a value $h = p$ such that the sum

$$G^{(p)}(u) = \sum_{j=1}^{p} G_j(u) \tag{III.3.13}$$

is (in view of what was said above) a solution of the system (III.2.11):

$$G(u + 2\pi i e^{(n)}) - G(u) = H_n(u), \qquad n = 1, 2, \dots, p.$$

The general solution of this system obviously has the form

$$G(u) = G^{(p)}(u) + P(u) = \sum_{j=1}^{p} G_j(u) + P(u), \tag{III.3.14}$$

where $P(u)$ is any entire function which has periods $2\pi i e^{(1)}, \dots, 2\pi i e^{(p)}$ (in particular, any constant).

§4. Jacobian or intermediate functions. The second system of difference equations

In the preceding two sections we showed that, starting from any reduced representation of an Abelian function $f(u) = \varphi(u)/\psi(u)$, one can obtain another: $f(u) = \varphi^*(u)/\psi^*(u)$, where $\varphi^*(u) = e^{G(u)}\varphi(u)$ and $\psi^*(u) = e^{G(u)}\psi(u)$ and where, moreover, $G(u)$ satisfies *the first system of difference equations* (III.2.11). This transition guarantees a sufficiently simple behavior of the functions $\varphi^*(u)$ and $\psi^*(u)$ with respect to the first p periods of the Riemann matrix $\Omega = (2\pi i E_p, T)$. In fact, when $2\pi i e^{(h)}$ $(h = 1, \dots, p)$ is added to u, the functions $\varphi^*(u)$ and $\psi^*(u)$ are multiplied by a factor of the form $\exp[l_h(u)]$, where $l_h(u)$ is *a linear function* $(h = 1, \dots, p)$. Since this property holds for *any* choice of a periodic entire function $P(u)$ in the formula (III.3.14) for $G(u)$, it suggests the idea of *making a particular choice* which would allow one to find a representation for $f(u)$ for which $\varphi^*(u)$ and $\psi^*(u)$ have equally simple properties with respect to *the remaining p periods of the matrix Ω*.

Assume that we specify the function $P(u)$ in the formula (III.3.13); for example, that we set $P(u) = 0$. We omit the star (*) in the expressions $\varphi^*(u) = \exp[G^{(p)}(u)]\varphi(u)$ and $\psi^*(u) = \exp[G^{(p)}(u)]\psi(u)$, considering the representation

$$f(u) = \frac{\varphi(u)}{\psi(u)} \tag{III.4.1}$$

as the starting point. Thus (III.4.1) satisfies the conditions

$$\varphi(u+2\pi i e^{(h)}) = \exp[l_h(u)]\varphi(u), \ \psi(u+2\pi i e^{(h)}) = \exp[l_h(u)]\psi(u), \quad \text{(III.4.2)}$$

where $l_h(u)$ $(h = 1, \dots, p)$ are linear functions whose coefficients do not depend on the particular choice of $P(u)$ (see III.2.10). As far as concerns the periods $t^{(k)}$ which are expressed by the columns of the matrix T, we can repeat the arguments of section III.2 for them with a somewhat altered notation; then we get

$$\varphi(u+t^{(k)}) = e^{m_k(u)}\varphi(u), \qquad \psi(u+t^{(k)}) = e^{m_k(u)}\psi(u), \qquad \text{(III.4.3)}$$

where $m_k(u)$ are entire functions determined by these equalities up to integer multiples of $2\pi i$. We now pass from $P \equiv 0$ to a periodic entire function $P(u)$ in the formula (III.3.14). Similarly, we pass from (III.4.1) to the representation

$$f(u) = \frac{\varphi^*(u)}{\psi^*(u)}, \qquad \text{(III.4.1}')$$

where

$$\varphi^*(u) = \exp[P(u)]\varphi(u), \qquad \psi^*(u) = \exp[P(u)]\psi(u). \qquad \text{(III.4.4)}$$

Then (III.4.2) will also hold for the functions $\varphi^*(u)$, $\psi^*(u)$, and with *the same linear functions* $l_h(u)$:

$$\varphi^*(u+2\pi i e^{(h)}) = \exp[l_h(u)]\varphi^*(u),$$
$$\psi^*(u+2\pi i e^{(h)}) = \exp[l_h(u)]\psi^*(u), \qquad \text{(III.4.2}')$$

whereas the functions $m_k(u)$ are transformed to new entire functions $m_k^*(u)$:

$$\varphi^*(u+t^{(k)}) = \exp[m_k^*(u)]\varphi^*(u), \quad \psi^*(u+t^{(k)}) = \exp[m_k^*(u)]\psi^*(u). \ \text{(III.4.5)}$$

From now on we will consider only the behavior of the function $\varphi^*(u)$.

We state the following problem: *to find an entire function* $P(u)$ *with periods* $2\pi i e^{(1)}, \dots, 2\pi i e^{(p)}$ *such that the corresponding functions* $m_k^*(u)$ *which have the form* (III.4.5) *are polynomials of degree no greater than* 1 (*and in general nonhomogeneous*):

$$m_k^*(u) = \sum_{s=1}^{p} \mu_{ks} u_s + b_k = \mu_k(u) + b_k, \qquad k = 1, \dots, p, \qquad \text{(III.4.6)}$$

where $\mu_k(u)$ *is a linear function and* b_k *is a constant.* If the problem is solvable, then $\varphi^*(u) = \exp[P(u)]\varphi(u)$ is an entire function which is multiplied by a factor of the form $\exp[g_k(u)]$, where $g_k(u)$ is a polynomial of degree not greater than 1 when we add to u any period $\omega^{(k)}$ $(k = 1, 2, \dots, 2p)$ given by the kth column of the matrix Ω.

In general, let Ω be any Riemann matrix with basic periods $\omega^{(k)}$ (which are not necessarily of the form $(2\pi i E_p, T)$) and let $\varphi(u)$ be an entire function satisfying the condition

$$\varphi(u+\omega^{(k)}) = \exp[\lambda_k(u) + c_k]\varphi(u), \qquad k = 1, \dots, 2p, \qquad \text{(III.4.7)}$$

where $\lambda_k(u)$ is a linear function and c_k is a constant (the latter is determined by the formula (III.4.7) only up to an integer multiple of $2\pi i$). Then we obviously have

$$\begin{aligned}\varphi(u+\omega^{(k)}+\omega^{(h)}) &= \exp[\lambda_k(u+\omega^{(h)})+c_k]\varphi(u+\omega^{(h)})\\ &= \exp[\lambda_k(u+\omega^{(h)})+\lambda_h(u)+c_k+c_h]\varphi(u)\\ &= \exp\{[\lambda_k(u)+\lambda_h(u)]+c_k+c_h+\lambda_k(\omega^{(h)})\}\varphi(u);\end{aligned} \tag{III.4.8}$$

and from this it follows, in particular, that for any natural number n

$$\varphi(u+n\omega^{(k)}) = \exp\left[n\lambda_k(u)+nc_k+\frac{1}{2}n(n-1)\lambda_k(\omega^{(k)})\right]\varphi(u) \tag{III.4.8$'$}$$

(this can be verified by induction). Furthermore (changing u in (III.4.7) into $u-\omega^{(k)}$), we get

$$\varphi(u) = \exp[\lambda_k(u-\omega^{(k)})+c_k]\varphi(u-\omega^{(k)}),$$

from which it follows that

$$\varphi(u-\omega^{(k)}) = \exp[-\lambda_k(u)-c_k+\lambda_k(\omega^{(k)})]\varphi(u). \tag{III.4.8$''$}$$

Since *any period* ω generated by the fundamental matrix Ω can be represented as a sum of terms of the form $\pm\,\omega^{(k)}$ (repeated several times), it follows from the previous remarks that there exist a linear function $\lambda(u)=\lambda_\omega(u)$ and a constant $c=c_\omega$ (determined up to an integer multiple of $2\pi i$), corresponding to the period, such that

$$\varphi(u+\omega) = \exp[\lambda_\omega(u)+c_\omega]\varphi(u). \tag{III.4.9}$$

Furthermore, if u is transformed by a nonsingular linear transformation $u^*=Au$ (as a result of which, to each period ω, there corresponds a period $\omega^*=A\omega$), then we have for the transformed function $\varphi^*(u^*)=\varphi(A^{-1}u^*)=\varphi(u)$:

$$\begin{aligned}\varphi^*(u^*+\omega^*) &= \varphi(u+\omega) = \exp[\lambda(u)+c]\varphi(u)\\ &= \exp[\lambda(A^{-1}u^*)+c]\varphi^*(u^*) = \exp[\lambda^*(u^*)+c^*]\varphi^*(u),\end{aligned} \tag{III.4.10}$$

where $\lambda^*(u^*)=\lambda(A^{-1}u^*)=\lambda(u)$ is a linear function of u^* and $c^*=c$. Therefore the properties expressed by the equalities (III.4.7) are still valid (with possibly different functions λ_k and constants c_k) when one changes from the matrix Ω to an equivalent matrix $\Omega^*=A\Omega M$ (section II.6; recall that the transformation ΩM, where M is a unimodular matrix, transforms one fundamental system of periods into another). This justifies the following definition.

An entire function $\varphi(u)$ is called *a Jacobian* or *intermediate function with respect to some equivalence class represented by a Riemann matrix* Ω if for any period ω generated by this matrix the following condition holds:

$$\varphi(u+\omega) = \exp[\lambda(u) + c]\varphi(u)\,, \qquad \text{(III.4.11)}$$

where $\lambda(u) = \lambda_\omega(u)$ is a linear function and $c = c_\omega$ is a constant.

If $Q(u)$ is any polynomial of degree not greater than two,

$$Q(u) = \sum_{j,k=1}^{p} a_{jk}u_j u_k + \sum_{l=1}^{p} b_l u_l + c\,,$$

then, independently of which particular p-dimensional complex vector ω is taken, the difference $Q(u+\omega) - Q(u) = \lambda_\omega(u)$ is a polynomial of degree not exceeding one. Hence

$$\exp[Q(u+\omega)] = \exp[\lambda(u) + c]\exp[Q(u)]\,,$$

i.e., *a function of the form* $\exp[Q(u)]$ *is a Jacobian function for any period matrix* Ω. However, our definition does not exclude degenerate functions; in particular, constants.

It follows from this definition that *the product of any two Jacobian functions corresponding to a matrix* Ω *is also a Jacobian function corresponding to the same matrix.*

Indeed, if

$$\varphi_1(u+\omega) = \exp[\lambda_1(u) + c_1]\varphi_1(u)\,,$$
$$\varphi_2(u+\omega) = \exp[\lambda_2(u) + c_2]\varphi_2(u)\,,$$

then

$$\varphi_1(u+\omega)\varphi_2(u+\omega) = \exp[\lambda_1(u) + \lambda_2(u) + c_1 + c_2]\varphi_1(u)\varphi_2(u)\,.$$

For any point $a \in \mathbf{C}^p$ *the functions* $\varphi(u+a)$ *and* $\varphi(u)$ *are both Jacobian functions.*

Indeed, if $\varphi(u+\omega) = \exp[\lambda(u) + c]\varphi(u)$, then

$$\varphi(u+a+\omega) = \exp[\lambda(u) + \lambda(a) + c]\varphi(u+a)\,.$$

We see that when *one moves from* $\varphi(u)$ *to* $\varphi(u+a)$ *only the constant term in the exponential changes by the value* $\lambda(a)$.

We will show later (section III.5) that any Abelian function can be expressed in the form of a quotient of two Jacobian functions. Note that no Jacobian function can be Abelian. More precisely, no entire function can be Abelian.

Indeed, if an entire function $\varphi(u)$ (nondegenerate, this time) has periods $2\pi i e^{(h)}$, $h = 1, \dots, p$, then since they are complex independent, it cannot have any integer period independent of them (section II.7) and therefore it cannot be a $2p$-periodic function.

In section II.4 we introduced an example of a Riemann matrix

$$\Omega = \begin{pmatrix} 1 & 2\tau & \cdots & 0 & 0 \\ 0 & 0 & \cdots & 0 & 0 \\ 0 & 0 & \cdots & 1 & 2\tau \end{pmatrix}, \quad \text{where } \operatorname{Im}\tau > 0.$$

By changing the order of the columns (which is accomplished by the simplest unimodular transformations), this matrix can be written in the form

$$\Omega = \begin{pmatrix} 1 & 0 & \cdots & 0 & 2\tau & 0 & \cdots & 0 \\ 0 & 1 & \cdots & 0 & 0 & 2\tau & \cdots & 0 \\ \cdots & \cdots & \cdots & \cdots & \cdots & \cdots & \cdots & \cdots \\ 0 & 0 & \cdots & 1 & 0 & 0 & \cdots & 2\tau \end{pmatrix},$$

which differs only insignificantly from the one considered in this chapter, i.e., from (III.1.11). There we also gave an example of the corresponding Abelian function

$$f(u) = \sum_{j=1}^{p} A_j \frac{\vartheta_3(u_j)}{\vartheta_4(u_j)}, \quad \text{where } A_j \neq 0,\ j = 1, \dots, p.$$

Writing it in the form

$$f(u) = \frac{A_1 \vartheta_3(u_1)\vartheta_4(u_2)\cdots\vartheta_4(u_p) + \cdots + A_p\vartheta_3(u_p)\vartheta_4(u_1)\cdots\vartheta_4(u_{p-1})}{\vartheta_4(u_1)\cdots\vartheta_4(u_p)},$$

we easily verify with the help of Table 1 (section I.3) that the numerator $\varphi(u)$ and the denominator $\psi(u)$ are Jacobian functions corresponding to the matrix Ω. Indeed, it follows from Table 1 that the functions $\vartheta_3(u_k)$ and $\vartheta_4(u_k)$ satisfy the relations

$$\begin{gathered} \vartheta_3(u_k + 1) = \vartheta_3(u_k), \qquad \vartheta_4(u_k + 1) = \vartheta_4(u_k), \\ \vartheta_3(u_k + 2\tau) = \exp[-4\pi i(u_k + \tau)]\vartheta_3(u_k), \\ \vartheta_4(u_k + 2\tau) = \exp[-4\pi i(u_k + \tau)]\vartheta_4(u_k). \end{gathered}$$

Hence

$$\begin{gathered} \varphi(u + e^{(k)}) = \varphi(u), \qquad \psi(u + e^{(k)}) = \psi(u), \\ \varphi(u + 2\tau e^{(k)}) = \exp[-4\pi i(u_k + \tau)]\varphi(u), \\ \psi(u + 2\tau e^{(k)}) = \exp[-4\pi i(u_k + \tau)]\psi(u). \end{gathered}$$

This means that $\varphi(u)$ and $\psi(u)$ are Jacobian functions.

Let us return to the general case. First, we will give an argument which is very similar to the arguments in section III.2. The only difference will be that we will now consider the behavior of the functions $m_k(u)$ induced by the relations (III.4.3) (or the behavior of the functions $m_k^*(u)$ induced by the relations (III.4.5)) with respect to two groups of periods $t^{(1)}, \dots, t^{(p)}$ and

$2\pi i e^{(1)}, \ldots, 2\pi i e^{(p)}$. Thus, with the help of (III.4.2) and (III.4.3), we get

$$\varphi(u+2\pi i e^{(h)}) = \exp\left\{-\frac{1}{2}\sum_{j=1}^{p} n_{hj}u_j\right\}\varphi(u),$$

$$\begin{aligned}\varphi(u+2\pi i e^{(h)}+t^{(k)}) &= \exp\left\{-\frac{1}{2}\sum_{j=1}^{p} n_{hj}(u_j+t_{jk})\right\}\varphi(u+t^{(k)})\\ &= \exp\left\{-\frac{1}{2}\sum n_{hj}(u_j+t_{jk})+m_k(u)\right\}\varphi(u).\end{aligned}$$

If we now perform the calculations in a different order we get

$$\begin{aligned}\varphi(u+t^{(k)}) &= \exp[m_k(u)]\varphi(u),\\ \varphi(u+t^{(k)}+2\pi i e^{(h)}) &= \exp[m_k(u+2\pi i e^{(h)})]\varphi(u+2\pi i e^{(h)})\\ &= \exp\left[m_k(u+2\pi i e^{(h)})-\frac{1}{2}\sum_{j=1}^{p} n_{hj}u_j\right]\varphi(u).\end{aligned}$$

Comparing the preceding results, we conclude that

$$m_k(u+2\pi i e^{(h)}) - m_k(u) = -\frac{1}{2}\sum_{j=1}^{p} n_{hj}t_{jk} + 2\pi i l_{hk} \qquad (h, k = 1, 2, \ldots, p),$$

where l_{hk} are real integers.

If one uses matrix notation

$$N = (n_{hj}), \qquad \mathscr{L} = (l_{hk}) \tag{III.4.12}$$

(the matrix N is skew-symmetric, as was noted in the section III.2), then the preceding relations can be represented in the following form:

$$m_k(u+2\pi i e^{(h)}) - m_k(u) = -\frac{1}{2}\tilde{e}^{(h)}Nt^{(k)} + 2\pi i l_{hk}. \tag{III.4.13}$$

Here and later, the sign $\sim$ means taking the transposed matrix, which in this case means changing the columns of the unit matrix into rows.

If we use (III.4.2$'$) and (III.4.5) instead of (III.4.2) and (III.4.3), then we get the relations

$$m_k^*(u+2\pi i e^{(h)}) - m_k^*(u) = -\frac{1}{2}\tilde{e}^{(h)}Nt^{(k)} + 2\pi i l_{hk}^*, \tag{III.4.13$'$}$$

where the integers l_{hk}^* make up the matrix $\mathscr{L}^*$:

$$\mathscr{L}^* = (l_{hk}^*), \qquad h = 1, \ldots, p; \; k = 1, \ldots, p. \tag{III.4.14}$$

In the same way as for the relations (III.4.13) we can find additional relations which characterize the behavior of the functions $m_k(u)$ with respect to the periods $t^{(h)}$. Let us express $\varphi(u+t^{(h)}+t^{(k)})$ in terms of $\varphi(u)$ in two different ways, using (III.4.3). Comparing the results, we find that

$$[m_k(u+t^{(h)}) - m_k(u)] - [m_h(u+t^{(k)}) - m_h(u)] = 2\pi i r_{hk}, \tag{III.4.15}$$

where r_{hk} are real integers making up a (p, p) matrix R:

$$R = (r_{hk}), \qquad h = 1, \dots, p; \; k = 1, \dots, p. \tag{III.4.16}$$

Similarly we have, for the functions $m^*(u)$ defined by the relations (III.4.5),

$$[m_k^*(u + t^{(h)}) - m_k^*(u)] - [m_k^*(u + t^{(k)}) - m_h^*(u)] = 2\pi i r_{hk}^*, \tag{III.4.15$'$}$$

where r_{hk}^* are also integers making up a (p, p) matrix R^*:

$$R^* = (r_{hk}^*), \qquad h = 1, \dots, p; \; k = 1, \dots, p. \tag{III.4.17}$$

Assume that for some entire function $P(u)$ with periods $2\pi i e^{(1)}, \dots, 2\pi i e^{(p)}$ the corresponding functions $m_k^*(u)$ have the form (III.4.6). We introduce a matrix of coefficients

$$M = (\mu_{ks}), \qquad k = 1, \dots, p; \; s = 1, \dots, p. \tag{III.4.18}$$

Moreover, let b be a p-vector whose coordinates b_k are the constant terms of the functions (III.4.6), that is,

$$\tilde{b} = (b_1, \dots, b_p). \tag{III.4.19}$$

Then (III.4.6) can be rewritten in the form

$$m_k^*(u) = \tilde{e}^{(k)}(Mu + b). \tag{III.4.6$'$}$$

Accordingly, the left-hand side of (III.4.13$'$) can be written as:

$$m_k^*(u + 2\pi i e^{(h)}) - m_k^*(u) = 2\pi i \tilde{e}^{(k)} M e^{(h)}.$$

The right-hand side of (III.4.13$'$) has (if one transposes the product of matrices $\tilde{e}^{(h)} N t^{(k)}$, which is equal to a (1,1)-matrix), the form

$$-\frac{1}{2}\tilde{t}^{(k)} \tilde{N} e^{(h)} + 2\pi i l_{hk}^*.$$

But N is a skew-symmetric matrix and hence $\tilde{N} = -N$ and moreover $\tilde{t}^{(k)} = \tilde{e}^{(k)} \tilde{T}$ and $l_{hk}^* = \tilde{e}^{(k)} \tilde{\mathscr{L}}^* e^{(h)}$. Consequently, the relation (III.4.13$'$) can be written in matrix form as

$$2\pi i \tilde{e}^{(k)} M e^{(h)} = \frac{1}{2}\tilde{e}^{(k)} \tilde{T} N e^{(h)} + 2\pi i \tilde{e}^{(k)} \tilde{\mathscr{L}}^* e^{(h)},$$

or

$$\tilde{e}^{(k)} \left[2\pi i M - \left(\frac{1}{2}\tilde{T} N + 2\pi i \tilde{\mathscr{L}}^*\right)\right] e^{(h)} = 0$$

for all $k = 1, \dots, p$ and $h = 1, \dots, p$. But this means that

$$M = \frac{1}{4\pi i}\tilde{T} N + \tilde{\mathscr{L}}^*. \tag{III.4.20}$$

Thus *the coefficients of linear functions* (except for components of the vector b) *can be expressed in terms of the entries of the complex (p, p) matrix T which forms the second part of the period matrix Ω, and entries of the integer matrices N and $\mathscr{L}^*$.*

Similarly, we can obtain a matrix equation corresponding to the relations (III.4.15′). With the help of (III.4.6′), the left-hand side of (III.4.15′) can be written in the form $\tilde{e}^{(k)}Mt^{(h)} - \tilde{e}^{(h)}Mt^{(k)}$; the right-hand side can be written as $2\pi i\tilde{e}^{(h)}R^*e^{(k)}$. Substituting $Te^{(h)}$ for $t^{(h)}$ and $Te^{(k)}$ for $t^{(k)}$ on the left-hand side, and transposing $\tilde{e}^{(k)}MTe^{(h)}$, we find that

$$\tilde{e}^{(h)}\widetilde{T}\widetilde{M}e^{(k)} - \tilde{e}^{(h)}MTe^{(k)} = 2\pi i\tilde{e}^{(h)}R^*e^{(k)},$$

from which we conclude, in the same way as above that

$$\widetilde{T}\widetilde{M} - MT = 2\pi iR^*. \tag{III.4.21}$$

Let us now turn to the conditions which the required function $P(u)$ must satisfy. From the first equations in (III.4.4) and (III.4.3) it follows that

$$\varphi^*(u + t^{(h)}) = \exp[P(u + t^{(h)}) - P(u) + m_h(u)]\varphi^*(u).$$

Let us compare this result with the first equation in (III.4.5) and set the corresponding integral multiple of $2\pi i$ equal to 0 (the function $m_k^*(u)$ is defined up to an integral multiple of $2\pi i$ in any case). We find

$$m_h^*(u) = P(u + t^{(h)}) - P(u) + m_h(u),$$

or

$$\begin{aligned} P(u + t^{(h)}) - P(u) &= m_h^*(u) - m_h(u) \\ &= H_h(u), \qquad h = 1, \dots, p. \end{aligned} \tag{III.4.22}$$

Thus *the function* $P(u)$ *must satisfy the system of difference equations* (III.4.22) that we call *the second system* to distinguish it from the system considered earlier. Then $m_h^*(u) - b_h$ are linear functions whose coefficient matrix M satisfies the equations (III.4.20) and (III.4.21) (so far, there are no restrictions on the constant terms b_h); and the functions $m_h(u)$ are defined by the equations (III.4.3).

§5. Solvability conditions for the second system and the solution of the second system

In order for an entire periodic function $P(u)$ satisfying the system (III.4.22) to exist (and therefore in order for the functions $m_k^*(u)$, which are generated by it, to exist) some conditions are needed. We now introduce them and verify that they hold if the matrix Ω is indeed a Riemann matrix.

First, in order for the system (III.4.22) to have a solution $P(u)$ with periods $2\pi ie^{(1)}, \dots, 2\pi ie^{(p)}$ it is necessary that the functions

$$H_k(u) = m_k^*(u) - m_k(u), \qquad k = 1, \dots, p, \tag{III.5.1}$$

have the same periods.

Consider the differences

$$\begin{aligned} H_k(u + 2\pi ie^{(h)}) - H_k(u) = [&m_k^*(u + 2\pi ie^{(h)}) - m_k^*(u)] \\ &- [m_k(u + 2\pi ie^{(n)}) - m_k(u)]. \end{aligned}$$

Because of the relations (III.4.13′) and (III.4.13), the differences can be written in the form

$$H_k(u+2\pi i e^{(h)}) - H_k(u) = 2\pi i(l^*_{hk} - l_{hk}). \tag{III.5.2}$$

Therefore *the first condition of solvability holds if and only if the matrices* $\mathscr{L}$ *and* $\mathscr{L}^*$ *are equal*, i.e., if and only if

$$\mathscr{L}^* = \mathscr{L}. \tag{III.5.3}$$

Because of the equality (III.4.20), this condition imposes constraints on the matrix M of coefficients of the functions $m^*_k(u)$. The validity of (III.4.20) requires that the following matrix equality is satisfied:

$$M = \frac{1}{4\pi i}\tilde{T}N + \widetilde{\mathscr{L}}. \tag{III.5.4}$$

Since the right-hand side of (III.5.4) contains fixed (p, p)-matrices T, N, and $\mathscr{L}$, the choice of the matrix M and therefore of the linear functions $\mu_k(u)$ as parts of $m^*_k(u)$ in (III.4.6) is complete. We still have the constants $b_1, \ldots, b_p$ at our disposal.

Turning to the derivation of further conditions for solvability, we add the equalities

$$P(u+t^{(h)}+t^{(k)}) - P(u+t^{(k)}) = H_h(u+t^{(k)}),$$
$$P(u+t^{(k)}) - P(u) = H_k(u);$$

and obtain

$$P(u+t^{(h)}+t^{(k)}) - P(u) = H_h(u+t^{(k)}) + H_k(u).$$

If we interchange the indices h and k and equate the right-hand sides, we obtain

$$H_{kh}(u) = [H_k(u+t^{(h)}) - H_k(u)] - [H_h(u+t^{(k)}) - H_h(u)] = 0, \qquad h, k = 1, \ldots, p. \tag{III.5.5}$$

This, then, is *the second set of conditions of solvability.* With the help of the formula (III.5.1), the expression $H_{kh}(u)$ can be represented in the form

$$\begin{aligned} H_{kh}(u) = {} & \{[m^*_k(u+t^{(h)}) - m^*_k(u)] - [m^*_h(u+t^{(k)}) - m^*_h(u)]\} \\ & - \{[m_k(u+t^{(h)}) - m_k(u)] - [m_h(u+t^{(k)}) - m_h(u)]\}, \end{aligned}$$

which, because of (III.4.15) and (III.4.15′), can be written as

$$H_{kh}(u) = 2\pi i(r^*_{hk} - r_{hk}),$$

where r^*_{hk} and r_{hk} are integers, which are the entries of the matrices R^* and R. Thus at this point, instead of the conditions (III.5.5) that are needed, we have the following weaker conditions:

$$H_{kh}(u) = 2\pi i(r^*_{hk} - r_{hk}), \qquad h, k = 1, \ldots, p. \tag{III.5.5′}$$

However, by the periodicity of the function $H_k(u)$, which is guaranteed by the choice of the matrix M in the formula (III.5.4), it is easy to show that the condition (III.5.5′) actually agrees completely with (III.5.5), i.e.,

$$R^* = R. \tag{III.5.6}$$

Indeed, each periodic entire function $H_k(u)$ can be represented, according to section II.7, as a generalized Fourier series

$$H_k(u) = \sum_{(\nu)} A_\nu^{(k)} e^{\tilde{\nu}\cdot u}, \qquad k = 1, \dots, p, \tag{III.5.7}$$

where the summation is over all the p-tuples of integers ν, which we consider as p-vectors, so that $\tilde{\nu} = (\nu_1, \dots, \nu_p)$, where $\nu_1, \dots, \nu_p$ are any integers. We write $H_k(u)$ in the form

$$H_k(u) = A_0^{(k)} + \sum_{\nu\neq 0} A_\nu^{(k)} e^{\tilde{\nu} u}, \qquad k = 1, \dots, p. \tag{III.5.8}$$

Now the function $H_{kh}(u)$ (the right-hand side of (III.5.5)) is represented as a Fourier series:

$$H_{kh}(u) = \sum_{\nu\neq 0} [A_\nu^{(k)}(e^{\tilde{\nu}t^{(h)}} - 1) - A_\nu^{(h)}(e^{\tilde{\nu}t^{(k)}} - 1)] e^{\tilde{\nu} u}. \tag{III.5.9}$$

The conditions (III.5.5′) show that H_{kh} does not depend on u, i.e., it is a constant. Hence all the coefficients of $e^{\tilde{\nu} u}$ $(\nu \neq 0)$ must be zero. In other words, the Fourier coefficients of the expansion (III.5.8) must be connected, because of the conditions (III.5.5′), by the relations

$$A_\nu^{(k)}(e^{\tilde{\nu}t^{(h)}} - 1) = A_\nu^{(h)}(e^{\tilde{\nu}t^{(k)}} - 1), \qquad k, h = 1, \dots, p, \tag{III.5.10}$$

for any $\nu \neq 0$. This, in turn, means that

$$H_{kh}(u) \equiv 0, \quad \text{i.e., } R^* = R,$$

and therefore conditions (III.5.5) are satisfied.

However *in order for the system* (III.4.22) *to be solvable* we must require the function $H_k(u)$ to satisfy *one more condition* (the last one). That is, if there exists an entire periodic function P satisfying this system, it must also have a Fourier series expansion

$$P(u) = \sum_{(\nu)} X_\nu e^{\tilde{\nu} u}. \tag{III.5.11}$$

Therefore, each of the equations in the system has the form

$$\sum_{\nu\neq 0} X_\nu (e^{\tilde{\nu}t^{(k)}} - 1) e^{\tilde{\nu} u} = A_0^{(k)} + \sum_{\nu\neq 0} A_\nu^{(k)} e^{\tilde{\nu} u}. \tag{III.5.12}$$

From this, it follows, by the uniqueness of the Fourier series, that

$$A_0^{(1)} = \cdots = A_0^{(p)} = 0. \tag{III.5.13}$$

In order to satisfy these conditions we will use constants $b_k = m_k^*(0)$ which we are still free to change. Since it follows from (III.5.1) and (III.5.8) that

$$H_k(0) = A_0^{(k)} + \sum_{\nu\neq 0} A_\nu^{(k)} = b_k - m_k(0),$$

i.e.,

$$A_0^{(k)} = b_k - m_k(0) - \sum_{\nu\neq 0} A_\nu^{(k)},$$

we set

$$b_k = m_k(0) + \sum_{\nu\neq 0} A_\nu^{(k)}, \qquad k = 1, \dots, p. \tag{III.5.14}$$

Then the conditions (III.5.13) will be satisfied.

At this point there is nothing to prevent us from solving the system of equations (III.4.22). By the preceding analysis, *the entire functions* $H_k(u)$ *which occur on the right-hand sides of the equations are represented by series of the form* (III.5.8) *whose coefficients satisfy the conditions* (III.5.12) *and* (III.5.13).

Let us consider the equalities (III.5.12) in greater detail. We will show that *if, for some* $\nu \neq 0$ *and* k, *where* $1 \le k \le p$, *the difference* $e^{\tilde{\nu} t^{(k)}} - 1$ *becomes zero, then the corresponding coefficient* $A_\nu^{(k)}$ *is also zero.* Assume that $A_\nu^{(k)} \neq 0$. By the lemma in section III.1, for each $\nu \neq 0$ there must be at least one nonzero difference in the set $e^{\tilde{\nu} t^{(1)}} - 1, \dots, e^{\tilde{\nu} t^{(p)}} - 1$. For example, let $e^{\tilde{\nu} t^{(h)}} - 1 \neq 0$ $(h \neq k)$. Then the left-hand side of the corresponding equality (III.5.10) is different from 0 for the values of the indices h and k under consideration, whereas the right-hand side is equal to zero ($e^{\tilde{\nu} t^{(k)}} - 1 = 0$), which is impossible.

Now the p equalities (III.5.10) corresponding to the same $\nu \neq 0$ can be written in the form of an equality among p fractions of the form

$$\frac{A_\nu^{(1)}}{e^{\tilde{\nu} t^{(1)}} - 1} = \cdots = \frac{A_\nu^{(h)}}{e^{\tilde{\nu} t^{(h)}} - 1} = \cdots = \frac{A_\nu^{(p)}}{e^{\tilde{\nu} t^{(p)}} - 1} \qquad (\nu \neq 0). \tag{III.5.15}$$

It follows from the lemma in section III.1 that among them there exists at least one fraction whose denominator is not zero. This value should then be assigned to all the fractions with denominator zero (together with the numerator). The equations (III.4.22), expressed in the form (III.5.12), give us

$$X_\nu (e^{\tilde{\nu} t^{(k)}} - 1) = A_\nu^{(k)}, \qquad k = 1, \dots, p, \tag{III.5.16}$$

for any $\nu \neq 0$. When $\nu = 0$ these equalities are also satisfied in view of (III.5.13); the coefficient X_0 remains undetermined. When $\nu \neq 0$ the value of X_ν can be written (taking (III.5.15) into account) in the form

$$X_\nu = \frac{A_\nu^{(1)}}{e^{\tilde{\nu} t^{(1)}} - 1} = \cdots = \frac{A_\nu^{(p)}}{e^{\tilde{\nu} t^{(p)}} - 1}. \tag{III.5.17}$$

Of course, only the fractions whose denominators are not zero have any meaning for us. Such fractions always exist.

Thus, *the function* $P(u)$ *we are looking for is determined (if it exists) uniquely, up to an additive constant* X_0. It remains to verify the convergence of the series (III.5.11) for the values of the coefficients X_ν that we have found. To do this, we need the lemma of section III.1 in its complete form. It follows from the inequality

$$\max_{1\le k\le p} |e^{\tilde{\nu}t^{(k)}} - 1| \ge \alpha > 0,$$

where $\nu \neq 0$ and α is independent of ν, that for some $h = h_\nu$, $1 \le h_\nu \le p$, $|e^{\tilde{\nu}t^{(h)}} - 1| \ge \alpha$. Hence

$$|X_\nu| = \left|\frac{A_\nu^{(h)}}{e^{\tilde{\nu}t^{(h)}} - 1}\right| \le \frac{1}{\alpha}|A_\nu^{(h)}| \le \frac{1}{\alpha}(|A_\nu^{(1)}| + \cdots + |A_\nu^{(p)}|).$$

But the series

$$\frac{1}{\alpha}\sum_{\nu\neq 0}(|A_\nu^{(1)}| + \cdots + |A_\nu^{(p)}|)|e^{\tilde{\nu}u}|,$$

which is obtained by adding the absolute values of the terms of each series (III.5.8), converges uniformly in each p-disk $|u_j| \le R$ $(j = 1, \dots, p)$. Consequently the series (III.5.11) converges absolutely and uniformly over such p-disks. Therefore it determines an entire function that satisfies the system of difference equations that was introduced above.

§6. First and second period matrices. The characteristic matrix N

We have proved that *an Abelian function* $f(u)$ *can always be represented in the form of a quotient of two Jacobian functions*

$$f(u) = \frac{\varphi(u)}{\psi(u)}.$$

This representation is said to be *reduced* since it is obtained from another reduced representation by multiplying the numerator and the denominator of the fraction by an entire function $e^{P(u)}$ with no zeros.

Note that in the proof we used only the property that $f(u)$ is a meromorphic function which has periods generated by Ω. Hence *the result found above is also valid for functions which have additional periods and, in particular, for degenerate meromorphic periodic functions.*

Let

$$\Omega = (\omega_{jk}), \qquad j = 1, \dots, p; \ k = 1, \dots, 2p, \tag{III.6.1}$$

be a Riemann matrix of the function $f(u)$; we will no longer assume that it is of any special type. Thus the $2p$ periods of the fundamental system $\omega^{(k)}$ $(k = 1, \dots, 2p)$ will not, at this time, be broken into two groups as they have been up to now. Because $\varphi(u)$ is a Jacobian function we have

$$\varphi(u + \omega^{(k)}) = \exp[l_k(u)]\varphi(u), \qquad k = 1, \dots, 2p, \tag{III.6.2}$$

where it will be useful to represent the functions $l_k(u)$ in the following form:

$$l_k(u) = \lambda_k(u) + 2\pi i\gamma_k = 2\pi i\left(\sum_{j=1}^{p}\lambda_{jk}u_j + \gamma_k\right) = 2\pi i\tilde{\varepsilon}^{(k)}(\tilde{\Lambda}u + \gamma). \quad \text{(III.6.3)}$$

Here Λ is a $(p, 2p)$ matrix

$$\Lambda = (\lambda_{jk}), \qquad j = 1, \dots, p, \; k = 1, \dots, 2p, \quad \text{(III.6.4)}$$

γ is a vector

$$\tilde{\gamma} = (\gamma_1, \dots, \gamma_{2p}), \quad \text{(III.6.5)}$$

and $\varepsilon^{(k)}$ is a vector given by the kth column of the $(2p, 2p)$ unit matrix E_{2p}.

The following terminology is used in the theory of Abelian functions. The entries of the matrix Ω are called *periods of the first kind* and the entries of Λ are called *periods of the second kind.* The corresponding matrices Ω and Λ are called *the first and second period matrices.* A vector γ which has $2p$ complex components is called *a parametric vector of Jacobian functions.*

Thus Ω, Λ, and γ completely characterize the behavior of Jacobian functions. This behavior is described by the formulas (III.6.2). Two Jacobian functions $\varphi(u)$ and $\psi(u)$ for which Ω, Λ, and γ are identical are called *functions of the same type*, or *equivariant functions* (gleichändrige Funktionen). The type to which the Jacobian functions belong will be denoted by a triple $\{\Omega, \Lambda, \gamma\}$.

Of course, there are relations among the matrices of periods Ω and Λ. In essence, we saw these relations in the preceding section for the case when Ω had the form (III.2.1). In order to derive these relations in the present section we apply a well-known method: we express $\varphi(u+\omega^{(h)}+\omega^{(k)})$ in two different ways by using the formulas (III.6.2). We get

$$\begin{aligned}\varphi(u + \omega^{(h)} + \omega^{(k)}) &= \exp[l_k(u + \omega^{(h)}) + l_h(u)]\varphi(u)\\ &= \exp[l_h(u + \omega^{(k)}) + l_k(u)]\varphi(u),\end{aligned}$$

from which we conclude, by comparing exponents and substituting the expressions (III.6.3) for $\varphi(u)$, $l_k(u)$ and $l_h(u)$, that

$$2\pi i(\tilde{\varepsilon}^{(k)}\tilde{\Lambda}\omega^{(h)} - \tilde{\varepsilon}^{(h)}\tilde{\Lambda}\omega^{(k)}) = 2\pi i n_{hk}, \quad \text{(III.6.6)}$$

where n_{hk} are integers. It follows from this that, in the first place, the $(2p, 2p)$ matrix

$$N = (n_{hk}), \qquad h = 1, \dots, 2p, \; k = 1, \dots, 2p, \quad \text{(III.6.7)}$$

is *skew-symmetric.* It is called *the characteristic matrix of the Jacobian functions $\varphi(u)$ and $\psi(u)$.* The relations (III.6.6) can be represented as a matrix equation which expresses N in terms of Ω and Λ. Notice that $\omega^{(h)} = \Omega\varepsilon^{(h)}$ and $\omega^{(k)} = \Omega\varepsilon^{(k)}$, and hence

$$\tilde{\varepsilon}^{(k)}\tilde{\Lambda}\Omega\varepsilon^{(h)} - \tilde{\varepsilon}^{(h)}\tilde{\Lambda}\Omega\varepsilon^{(k)} = n_{hk};$$

but $\tilde{\varepsilon}^{(k)}\tilde{\Lambda}\Omega\varepsilon^{(h)} = \tilde{\varepsilon}^{(h)}\tilde{\Omega}\Lambda\varepsilon^{(k)}$ and therefore

$$\tilde{\varepsilon}^{(h)}(\tilde{\Omega}\Lambda - \tilde{\Lambda}\Omega)\varepsilon^{(k)} = n_{hk}, \qquad h = 1, 2, \dots, 2p, k = 1, \dots, 2p,$$

from which it finally follows that

$$\tilde{\Omega}\Lambda - \tilde{\Lambda}\Omega = N. \tag{III.6.8}$$

This, then, is the relation we were looking for. We stress that even though all the entries of the matrices Ω and Λ are complex numbers, the elements of the characteristic matrix N are real numbers and, moreover, integers.

Let us also consider the $(2p, 2p)$ matrix $\binom{\Omega}{\Lambda}$ and set $\begin{vmatrix}\Omega\\ \Lambda\end{vmatrix} = \delta$. The absolute value $|\delta|$ of this determinant, which (as we will show later) is a nonzero integer, plays an important role in the theory of Abelian functions and is called the *order* of the functions $\varphi(u)$ and $\psi(u)$ (Frobenius) or their *determinant* (Castelnuovo). Let us first show that δ is an integer. It follows from the formula (III.6.8) that N is the product of the two $(2p, 2p)$ matrices:

$$N = (\tilde{\Omega}, -\tilde{\Lambda})\begin{pmatrix}\Lambda\\ \Omega\end{pmatrix},$$

from which it follows that $|N| = |\tilde{\Omega}, -\tilde{\Lambda}|\begin{vmatrix}\Lambda\\ \Omega\end{vmatrix}$. But

$$|\tilde{\Omega}, -\tilde{\Lambda}| = \begin{vmatrix}\Omega\\ -\Lambda\end{vmatrix} = (-1)^p\begin{vmatrix}\Omega\\ \Lambda\end{vmatrix} = (-1)^p\delta$$

and

$$\begin{vmatrix}\Lambda\\ \Omega\end{vmatrix} = (-1)^p\begin{vmatrix}\Omega\\ \Lambda\end{vmatrix} = (-1)^p\delta.$$

Therefore

$$|N| = \delta^2. \tag{III.6.9}$$

Since N is a skew-symmetric matrix of even order $2p$ its determinant can be represented as the square of *the Pfaffian*, which is an entire rational function of the entries of N with integer coefficients (see section A.1 of the Appendix). Thus $|N| = P^2$, where P is an *integer*, since the elements of N are integers. Comparing this with (III.6.9) we conclude that $\delta = \pm P$ is also an integer. Thus the "determinant" $|\delta|$ is always a nonnegative integer.

Before we prove that $\delta \neq 0$ and therefore $|\delta| \geq 1$, we will prove that δ *is invariant with respect to the transition from the Riemann matrix* Ω *to a matrix* Ω^* *which is equivalent to* Ω:

$$\Omega^* = A\Omega M \tag{III.6.10}$$

(see section II.6). We will discover along the way how these linear transformations change the second period matrix Λ and the characteristic matrix N.

Consider the transformations M and A separately. It was shown above that we have the formulas (III.4.8), (III.4.8′) and (III.4.8″) for the functions $\varphi(u)$ which satisfy conditions of the form (III.4.7). Applying them repeatedly we find that for each period ω generated by a matrix Ω of the form

$$\omega = m_1\omega^{(1)} + \cdots + m_{2p}\omega^{(2p)},$$

the formula $\varphi(u+\omega) = \exp[\lambda_\omega(u)+c_\omega]\varphi(u)$ is valid where $\lambda_\omega(u) = m_1\lambda_1(u) + \cdots + m_{2p}\lambda_{2p}(u)$ and c_ω is a constant determined by the integers m_j $(j = 1, \ldots, 2p)$.

We now move from the fundamental system $\Omega = (\omega^{(1)}, \ldots, \omega^{(2p)})$ to another fundamental system $\Omega^* = (\omega^{(1)*}, \ldots, \omega^{(2p)*})$ (with the same variables $u_1, \ldots, u_p$). Then $\Omega^* = \Omega M$, where $M = (m_{jk})$, $j = 1, \ldots, 2p$; $k = 1, \ldots, 2p$ is a unimodular matrix (section II.6). In view of the preceding remarks, the following function corresponds in the new system to a period $\omega^{(k)*} = m_{1,k}\omega^{(1)} + \cdots + m_{2p,k}\omega^{(2p)}$:

$$\lambda_k^*(u) = m_{1,k}\lambda_1(u) + \cdots + m_{2p,k}\lambda_{2p}(u).$$

But this means that if $\lambda_k(u) = 2\pi i\tilde{\varepsilon}^{(k)}\tilde{\Lambda}u$, then $\lambda_k^*(u) = 2\pi i\tilde{\varepsilon}^{(k)}\tilde{\Lambda}^*u$, where $\Lambda^* = \Lambda M$.

Thus *as the result of a unimodular transformation both matrices* Ω *and* Λ *are transformed according to the same law,*

$$\Omega^* = \Omega M, \qquad \Lambda^* = \Lambda M. \tag{III.6.11}$$

From this, by formula (III.6.8), we obtain the following formula for the new characteristic matrix N^*

$$N^* = \tilde{\Omega}^*\Lambda^* - \tilde{\Lambda}^*\Omega^* = \tilde{M}\tilde{\Omega}\Lambda M - \tilde{M}\tilde{\Lambda}\Omega M = \tilde{M}NM. \tag{III.6.12}$$

Therefore

$$\delta^{*^2} = |N^*| = |\tilde{M}|\,|N|\,|M| = |N| = \delta^2$$

because $|M| = \pm 1$. Of course it follows from this that

$$\delta^* = \begin{vmatrix} \Omega^* \\ \Lambda^* \end{vmatrix} = \pm\delta.$$

This conclusion can, however, be obtained by a direct calculation:

$$\delta^* = \begin{vmatrix} \Omega^* \\ \Lambda^* \end{vmatrix} = \begin{vmatrix} \Omega M \\ \Lambda M \end{vmatrix} = \begin{vmatrix} \Omega \\ \Lambda \end{vmatrix} |M| = \pm\delta.$$

That is,

$$|\delta^*| = |\delta|. \tag{III.6.13}$$

Assume now that the variables $u_1, \dots, u_p$ are acted on by the transformation

$$u^* = Au, \qquad \text{(III.6.13)}$$

where $|A| \neq 0$. Then as we know (section II.6),

$$\Omega^* = A\Omega. \qquad \text{(III.6.15)}$$

In section III.5 we noticed that if

$$\varphi(u+\omega) = \exp[\lambda(\omega) + c]\varphi(u),$$

then for the transformed function $\varphi^*(u^*) = \varphi(u)$ and for any period $\omega^* = A\omega$ of this function, relation (III.4.10) holds:

$$\varphi^*(u^* + \omega^*) = \exp[\lambda^*(u^*) + c^*]\varphi^*(u^*),$$

where $\lambda^*(u^*) = \lambda(A^{-1}u^*) = \lambda(u)$ and $c^* = c$. Hence if $\lambda_k(u) = 2\pi i \tilde{\varepsilon}^{(k)}\widetilde{\Lambda}u$, the corresponding function $\lambda_k^*(u^*)$ has the form

$$\lambda_k^*(u^*) = \lambda_k(A^{-1}u^*) = 2\pi i \tilde{\varepsilon}^{(k)}\widetilde{\Lambda}A^{-1}u^*.$$

This means that the second period matrix Λ^* which is obtained as the result of the transformation satisfies the equality

$$\widetilde{\Lambda}^* = \widetilde{\Lambda}A^{-1}, \quad \text{i.e., } \Lambda^* = \widetilde{A}^{-1}\Lambda. \qquad \text{(III.6.16)}$$

From this, we obtain

$$\begin{aligned} N^* &= \widetilde{\Omega}^*\Lambda^* - \widetilde{\Lambda}^*\Omega^* = \widetilde{\Omega}\widetilde{A}\widetilde{A}^{-1}\Lambda - \widetilde{\Lambda}A^{-1}A\Omega \\ &= \widetilde{\Omega}\Lambda - \widetilde{\Lambda}\Omega = N \end{aligned} \qquad \text{(III.6.17)}$$

for the characteristic matrix N^*. Hence $|N^*| = |N| = \delta^2$, i.e., $|\delta^*| = |\delta|$ again. However, we can see that this is true by a direct calculation:

$$\delta^* = \begin{vmatrix} \Omega^* \\ \Lambda^* \end{vmatrix} = \begin{vmatrix} A\Omega \\ \widetilde{A}^{-1}\Omega \end{vmatrix} = \begin{vmatrix} A & 0 \\ 0 & \widetilde{A}^{-1} \end{vmatrix} \begin{vmatrix} \Omega \\ \Lambda \end{vmatrix} = |A|\,|\widetilde{A}^{-1}|\delta = \delta.$$

§7. Upper bound on absolute values of Jacobian functions. Riemann inequality

We saw in section III.4 that for any polynomial $Q(u)$ of degree not greater than 2, the function $\exp[Q(u)]$ is a Jacobi function; therefore as long as a function $\varphi(u)$ is Jacobian the product $\exp[Q(u)]\varphi(u)$ is also a Jacobi function. The main task of this section is to prove the following statement:

THEOREM III.7.1. *For all Jacobi functions $\varphi(u)$ of the same type, i.e., with common Ω, Λ, and γ, there exists a Hermitian matrix H such that for some polynomial $Q(u)$ (the same for all functions of this type) the following inequality holds:*

$$|\exp[-2\pi i Q(u)]\varphi(u)| \le G\exp[\tilde{u}H\bar{u}], \tag{III.7.1}$$

where $G = G(\varphi)$ is a constant.

First recall that a matrix $H = (h_{jk})$, $j, k = 1, \ldots, p$, defining a quadratic form

$$\tilde{u}H\bar{u} = \sum h_{jk}u_j\bar{u}_k \tag{III.7.2}$$

(as well as the form itself), is called *Hermitian* if $\tilde{\overline{H}} = H$.

For a hermitian matrix H there always exists a nonsingular (p, p) matrix T such that we obtain, as a result of the transformation $u = Tv$,

$$\tilde{u}H\bar{u} = \tilde{v}\tilde{T}H\overline{T}\bar{v} = h_1v_1\bar{v}_1 + \cdots + h_pv_p\bar{v}_p, \tag{III.7.3}$$

where h_i has the values ± 1 or 0 ($i = 1, \ldots, p$).

A Hermitian matrix (form) is called *positive definite* if, for some transformation $u = Tv$ $(|T| \ne 0)$, all h_i in the form (III.7.3) are equal to 1, i.e., the form can be reduced to the sum of the squares of the absolute values of the p variables $v_1, \ldots, v_p$. Obviously, a positive definite form is completely characterized by the condition

$$\tilde{u}H\bar{u} > 0, \quad \text{if } u \ne 0. \tag{III.7.4}$$

Indeed, because of the definition and because of (III.7.3), $\tilde{u}H\bar{u} \ge 0$. Moreover it can become zero only when $v_1 = \cdots = v_p = 0$, i.e., $u = Tv = 0$. The condition (III.7.4) can be written concisely as $H > 0$.

Notice that H is positive definite at the same time as H^{-1}, $\overline{H}$, and $\tilde{H}$. Indeed, if all h_j in the formula (III.7.3) are equal to 1 for some T, this is equivalent to the equality

$$\tilde{T}H\overline{T} = E_p. \tag{III.7.5}$$

But then $H = \tilde{T}^{-1}\overline{T}^{-1}$, $H^{-1} = \overline{T}\tilde{T}$, and therefore if we set $T' = \tilde{\overline{T}}^{-1}$ we obtain

$$\tilde{T}'H^{-1}\overline{T}' = \overline{T}^{-1}(\overline{T}\tilde{T})\tilde{T}^{-1} = E_p.$$

We can verify this in the same way for $\tilde{H}$ and $\overline{H}$.

Finally, a Hermitian matrix H is called *negative definite* if $-H$ is positive definite. For a negative definite matrix H,

$$\tilde{u}H\bar{u} < 0, \quad \text{if } u \ne 0. \tag{III.7.4}$$

We now turn to the verification of Theorem III.7.1. It will follow from this that the form $\tilde{u}H\bar{u}$ must be *positive definite*. Otherwise it could be reduced to the form (III.7.3) by some transformation $u = Tv$, where at least one of the coefficients h_j, say h_p, is not positive: $h_p \le 0$. Specify the values of the variables $v_1, \ldots, v_{p-1}$ arbitrarily. Then $\exp[-2\pi i Q(u)]\varphi(u) = g(v_1, \ldots, v_p)$ will be an entire function of v_p, bounded in absolute value by (III.7.1), and therefore will be a constant. That is, this function is independent of v_p. We can say the same for $\exp[-2\pi i Q(u)]\psi(u)$. It follows that

$$f(u) = \frac{\exp[-2\pi i Q(u)]\varphi(u)}{\exp[-2\pi i Q(u)]\psi(u)}$$

is a degenerate function, contrary to the definition of an Abelian function.

We begin the proof of the theorem by proving two lemmas.

LEMMA 1. *Let z be a vector in the space $\mathbf{C}^{2p}$; consider a linear transformation*

$$w = \begin{pmatrix} \Omega \\ \overline{\Omega} \end{pmatrix} z = Fz, \tag{III.7.6}$$

where $\Omega = (\omega_{jk})$ $(j = 1, \ldots, p;\ k = 1, \ldots, 2p)$ is a Riemann matrix. Let w' denote the vector formed by the first p components of w and by w'' the vector formed by the rest of the components of w, i.e., $w = \binom{w'}{w''}$. Then we have the formulas

$$z = F_1 w' + \overline{F}_1 w'', \tag{III.7.7}$$

for the inverse transformation, where F_1 is a $(2p, p)$ matrix and in addition

$$\Omega F_1 = \overline{\Omega}\,\overline{F}_1 = E_p, \qquad \overline{\Omega} F_1 = \Omega \overline{F}_1 = 0. \tag{III.7.8}$$

PROOF. It was established in section III.1 that $|F| \neq 0$ (III.1.7). Hence (III.7.6) has an inverse F^{-1}:

$$z = F^{-1} w = F^{-1} \begin{pmatrix} w' \\ w'' \end{pmatrix}.$$

We can rewrite this in the form $F^{-1} = (F_1, F_2)$ where F_1 and F_2 are $(2p, p)$ matrices (F_1 consists of the first p columns of F^{-1}). Then

$$F^{-1}F = (F_1, F_2) \begin{pmatrix} \Omega \\ \overline{\Omega} \end{pmatrix} = F_1 \Omega + F_2 \overline{\Omega} = E_{2p},$$

from which it follows that

$$\overline{F}_2 \overline{\Omega} + \overline{F}_1 \Omega = E_{2p},$$

i.e., F^{-1} can also be represented in the form $F^{-1} = (\overline{F}_2, \overline{F}_1)$ (in other words, $F^{-1}F = E_{2p}$ still holds for such a representation). Hence $F_2 = \overline{F}_1$ and therefore

$$F^{-1} = (F_1, \overline{F}_1), \tag{III.7.9}$$

from which it follows that $z = (F_1, \overline{F}_1)\binom{w'}{w''} = F_1 w' + \overline{F}_1 w''$. This completes the proof of (III.7.7).

It remains only to notice that

$$E_{2p} = \begin{pmatrix} E_p, & 0 \\ 0, & E_p \end{pmatrix} = \begin{pmatrix} \Omega \\ \overline{\Omega} \end{pmatrix} (F_1, \overline{F}_1) = \begin{pmatrix} \Omega F_1, & \Omega \overline{F}_1 \\ \overline{\Omega} F_1, & \overline{\Omega}\,\overline{F}_1 \end{pmatrix},$$

from which it follows that $\Omega F_1 = \overline{\Omega}\,\overline{F}_1 = E_p$, $\Omega\overline{F}_1 = \overline{\Omega}F_1 = 0$, completing the proof of (III.7.8).

We apply (III.7.6) to the special case when the coordinates of the vector $z = x$ are real numbers: $z_j = x_j$ $(j = 1, \ldots, 2p)$; then (III.7.6) gives us a vector w of the space $\mathbf{C}^{2p}$. Here

$$w' = \Omega x = u, \qquad w'' = \overline{\Omega} x = \bar{u}, \tag{III.7.6$'$}$$

and therefore (III.7.6) can be written as

$$\begin{pmatrix} u \\ \bar{u} \end{pmatrix} = \begin{pmatrix} \Omega \\ \overline{\Omega} \end{pmatrix} x,$$

and (III.7.7) becomes

$$x = F_1 u + \overline{F}_1 \bar{u}. \tag{III.7.7$'$}$$

Before stating the next lemma, which contains the main idea of the theorem that we are proving, we introduce the matrix $S = (s_{jk})$ $(j, k = 1, \ldots, 2p)$ defined by the equality

$$S = \widetilde{\Omega}\Lambda + \widetilde{\Lambda}\Omega. \tag{III.7.10}$$

It is obvious that this matrix is symmetric and that it is related to the characteristic matrix N (III.6.8) by the relation

$$S - N = 2\widetilde{\Lambda}\Omega. \tag{III.7.11}$$

We also introduce the $2p$-vector s whose components are the diagonal elements of the matrix S:

$$\tilde{s} = (s_{1,1}, s_{2,2}, \ldots, s_{2p,2p}). \tag{III.7.12}$$

We will use S to construct a polynomial $\Phi(x)$ of degree two with respect to the real variables x_j, $j = 1, \ldots, 2p$ $(\tilde{x} = (x_1, \ldots, x_{2p}))$. If we define x by (III.7.7$'$), we can consider $\Phi(x)$ as *a function of p complex variables* $u_1, \ldots, u_p$ (not an analytic function). The success of the role which $\Phi(x)$ will play later depends completely on the fact that the differences of the form $2\pi i[\Phi(x + e^{(h)}) - \Phi(x)]$, $h = 1, \ldots, 2p$, coincide (as we shall show below) with the functions (III.6.3), up to of pure imaginary additive terms.

Thus, let

$$\Phi(x) = \frac{1}{4}\tilde{x}Sx + \tilde{a}x, \tag{III.7.13}$$

where a is the $2p$ vector defined by

$$a = \gamma + \frac{1}{4}s. \tag{III.7.14}$$

LEMMA 2. *For a function of p complex variables* $u_1, \ldots, u_p$:

$$p(u) = \varphi(u)\exp[-2\pi i \Phi(x)] \tag{III.7.15}$$

(not analytic with respect to them, but obviously continuous over $\mathbf{C}^p$*) there exists a positive real constant* $G = G(\varphi)$ *such that the following inequality holds over the whole space* $\mathbf{C}^p$ *:*

$$|p(u)| \le G. \tag{III.7.16}$$

It is enough to prove that the absolute value of the function (III.7.15) is *a periodic function* with periods $\omega^{(h)}$, $h = 1, \dots, 2p$. Then we can define G as the *upper bound* of $|p(u)|$ at the points of the parallelotope Π_0 generated by the periods $\omega^{(h)}$ of Ω:

$$G = \sup |p(u)|, \qquad u \in \Pi_0. \tag{III.7.17}$$

By the definition of $\Phi(x)$ we have, for any $\varepsilon^{(h)}$ of the hth column of the unit matrix E_{2p},

$$\begin{aligned} 2\pi i[\Phi(x + \varepsilon^{(h)}) - \Phi(x)] &= 2\pi i \left[\frac{1}{4}(\tilde{x} + \tilde{\varepsilon}^{(h)})S(x + \varepsilon^{(h)}) \right. \\ &\qquad \left. + \tilde{a}(x + \varepsilon^{(h)}) - \frac{1}{4}\tilde{x}Sx - \tilde{a}x\right] \\ &= 2\pi i \left(\frac{1}{2}\tilde{\varepsilon}^{(h)}Sx + \gamma_h\right) \end{aligned}$$

(here we used the symmetry of S and the equalities (III.7.14) and (III.7.12)). By (III.7.11) this result can be presented as

$$\begin{aligned} 2\pi i[\Phi(x + \varepsilon^{(h)}) - \Phi(x)] &= 2\pi i \left[\frac{1}{2}\tilde{\varepsilon}^{(h)}(S - N)x + \tilde{\varepsilon}^{(h)}\gamma + \frac{1}{2}\tilde{\varepsilon}^{(h)}Nx\right] \\ &= 2\pi i\tilde{\varepsilon}^{(h)}\left[\widetilde{\Lambda}\Omega x + \gamma + \frac{1}{2}Nx\right], \end{aligned}$$

and finally, because $u = \Omega x$ (III.7.6), we have

$$2\pi i[\Phi(x + \varepsilon^{(h)}) - \Phi(x)] = 2\pi i\tilde{\varepsilon}^{(h)}(\widetilde{\Lambda}u + \gamma) + \pi i\tilde{\varepsilon}^{(h)}Nx. \tag{III.7.18}$$

Here the right-hand side differs from the function (III.6.3), which occurs in the exponential function on the right of (III.6.2), only by a pure imaginary term $\pi i\tilde{\varepsilon}^{(h)}Nx$. The function $\Phi(x)$ was introduced for exactly this purpose.

Because $\Omega x = u$, we have $\Omega(x + \varepsilon^{(h)}) = u + \omega^{(h)}$. Since these relations are invertible (if x is a real $2p$-vector), replacing u by $\omega^{(h)}$ corresponds to replacing x by $\varepsilon^{(h)}$, and we therefore have

$$p(u + \omega^{(h)}) = \varphi(u + \omega^{(h)})\exp[-2\pi i\Phi(x + \varepsilon^{(h)})].$$

But $\varphi(u + \omega^{(h)}) = \varphi(u)\exp[-2\pi i\tilde{\varepsilon}^{(h)}(\widetilde{\Lambda}u + \gamma)]$, and hence $p(u + \omega^{(h)}) = p(u)\exp[-2\pi i[\Phi(x+\varepsilon^{(h)})-\Phi(x)]+2\pi i\tilde{\varepsilon}^{(h)}(\widetilde{\Lambda}u+\gamma)] = p(u)\exp[-2\pi i\tilde{\varepsilon}^{(h)}Nx]$, from which it finally follows that

$$|p(u + \omega^{(h)})| = |p(u)|, \qquad h = 1, \dots, 2p. \tag{III.7.19}$$

Thus we have proved that $|p(u)|$ is periodic; hence Lemma 2 is established.

We turn now to the proof of the theorem. For this proof we must verify that, in the result

$$x = F_1 u + \overline{F}_1 \bar{u}$$

of the transformation (III.7.7′), the imaginary part of $\Phi(x)$ can be represented in the form

$$\operatorname{Im}\Phi(x) = \operatorname{Im}\left[Q(u) + \frac{1}{2\pi i}\tilde{u}H\bar{u}\right], \tag{III.7.20}$$

where $Q(u)$ is quadratic in the variables $u_1, \dots, u_p$, and H is a Hermitian matrix. In fact, by Lemma 2 we then obtain

$$\begin{aligned}|\varphi(u)\exp[-2\pi i Q(u)]| &= |p(u)\exp\{2\pi i[\Phi(x) - Q(u)]\}| \\ &\leq G\exp(\tilde{u}H\bar{u}).\end{aligned}$$

In order to obtain an expression of the form (III.7.20), we compute $\operatorname{Im}\Phi(x)$ under the condition $x = F_1 u + \overline{F}_1 \bar{u}$. Here we may replace $\Phi(x)$ by other suitable functions $\Phi_1(x), \Phi_2(x), \dots$, which differ from each other only by real terms.

Setting $a = a' + ia''$, where a' and a'' are vectors with real components, we have

$$\Phi_1(x) = \Phi(x) - \frac{1}{4}\tilde{x}Nx - \tilde{a}'x = \frac{1}{4}\tilde{x}(S-N)x + i\tilde{a}''x.$$

However, in view of (III.7.11), $\Phi_1(x) = \frac{1}{2}\tilde{x}\tilde{\Lambda}\Omega x + i\tilde{a}''x$. We now use (III.7.6′) and (III.7.7′) to obtain

$$\Phi_1(x) = \frac{1}{2}\tilde{u}\tilde{F}_1\tilde{\Lambda}u + \frac{1}{2}\tilde{u}\tilde{\overline{F}}_1\tilde{\Lambda}u + i\tilde{a}''(F_1 u + \overline{F}_1\bar{u});$$

in other words, by interchanging rows and columns in the matrices in the first two terms, we get

$$\Phi_1(x) = \frac{1}{2}\tilde{u}\Lambda F_1 u + \frac{1}{2}\tilde{u}\Lambda\tilde{\overline{F}}_1\bar{u} + ia''(F_1 u + \overline{F}_1\bar{u}).$$

Let us set

$$\frac{1}{2}\tilde{u}\Lambda F_1 u + 2i\tilde{a}''F_1 u = Q(u). \tag{III.7.21}$$

It is obvious that this is a quadratic polynomial in $u_1, \dots, u_p$, and it depends only on the matrices Ω, Λ, and on the vector γ (because the vector a does). On the other hand it does not depend on $\varphi(u)$. We have

$$\Phi_1(x) = Q(u) + \frac{1}{2}\tilde{u}\Lambda\overline{F}_1\bar{u} + ia''(\overline{F}_1\bar{u} - F_1 u);$$

discarding the last term, which is real, we obtain

$$\Phi_2(x) = Q(u) + \frac{1}{2}\tilde{u}\Lambda\overline{F}_1\bar{u}. \tag{III.7.22}$$

By Lemma 1, $\tilde{F}_1\tilde{\Omega} = \widetilde{\Omega F_1} = E_p$, $\Omega\overline{F}_1 = 0$, and therefore

$$\Lambda\overline{F}_1 = (\tilde{F}_1\tilde{\Omega})\Lambda\overline{F}_1 - \tilde{F}_1\tilde{\Lambda}(\Omega\overline{F}_1) = \tilde{F}_1(\tilde{\Omega}\Lambda - \tilde{\Lambda}\Omega)\overline{F}_1 = \tilde{F}_1 N\overline{F}_1. \quad \text{(III.7.23)}$$

Let us set

$$H = \pi i\Lambda\overline{F}_1 = \pi i\tilde{F}_1 N\overline{F}_1; \quad \text{(III.7.24)}$$

it is obvious that this is a Hermitian matrix, since

$$\overline{H} = -\pi i\overline{\tilde{F}}_1 N F_1 \quad \text{and} \quad \tilde{\overline{H}} = -\pi i\tilde{F}_1\tilde{N}\overline{F}_1 = \pi i\tilde{F}_1 N\overline{F}_1 = H.$$

The equality (III.7.20) follows from the preceding calculations,

$$\operatorname{Im}\Phi(x) = \operatorname{Im}\Phi_1(x) = \operatorname{Im}\Phi_2(x) = \operatorname{Im}\left[Q(u) + \frac{1}{2}\tilde{u}\Lambda\overline{F}_1\bar{u}\right]$$

$$= \operatorname{Im}\left[Q(u) + \frac{1}{2\pi i}\tilde{u}H\bar{u}\right];$$

this completes the proof of the theorem.

A corollary of this theorem, which was mentioned at the beginning of this section, deserves the status of a theorem.

THEOREM III.7.2 (Riemann inequality). *A Hermitian form*

$$\tilde{u}H\bar{u} = \pi i\tilde{u}\Lambda\overline{F}_1\bar{u} = \pi i\tilde{u}\tilde{F}_1 N\overline{F}_1\bar{u} \quad \text{(III.7.25)}$$

is positive definite.

From this theorem it is easy to obtain the following important result.

THEOREM III.7.3. *If Ω and Λ are the first and the second period matrices and $N = \tilde{\Omega}\Lambda - \tilde{\Lambda}\Omega$ is the corresponding characteristic matrix, we have*

$$|N| = \delta^2 \neq 0. \quad \text{(III.7.26)}$$

We first prove a lemma.

LEMMA 3. *If z is a complex $2p$-vector, not 0, and $\overline{\Omega}z = 0$, it follows that*

$$i\tilde{z}N\bar{z} > 0. \quad \text{(III.7.27)}$$

To prove the lemma, we apply the transformation (III.7.6) to the vector z and write the $2p$-vector w in the form $w = \binom{u}{w''}$, where w' $(= u)$ and w'' are p-vectors. We obtain $u = \Omega z$ and $w'' = \overline{\Omega}z = 0$. Hence the formula (III.7.7) takes the form $z = F_1 u$. Therefore formulas (III.7.23) and (III.7.24) imply

$$i\tilde{z}N\bar{z} = i\tilde{u}\tilde{F}_1 N\overline{F}_1\bar{u} = \pi\tilde{u}Hu,$$

from which it follows by Theorem III.7.2 that $i\tilde{z}N\bar{z} > 0$ if $z \neq 0$.

We now turn to the proof of Theorem III.7.3. Assume, on the contrary, that

$$\begin{vmatrix}\Omega \\ \Lambda\end{vmatrix} = \delta = 0.$$

Then we also have

$$\begin{vmatrix} \overline{\Omega} \\ \overline{\Lambda} \end{vmatrix} = \bar{\delta} = 0 .$$

Hence there is a $2p$ vector z, $z \neq 0$, for which the following system of equations is satisfied:

$$\overline{\Omega} z = 0 , \qquad \overline{\Lambda} z = 0 ,$$

and therefore $\Omega \bar{z} = 0$, $\Lambda \bar{z} = 0$. If we then replace N by its equivalent $\widetilde{\Omega}\Lambda - \widetilde{\Lambda}\Omega$, we obtain

$$i\tilde{z}N\bar{z} = i(\tilde{z}\widetilde{\Omega}\Lambda\bar{z} - \tilde{z}\widetilde{\Lambda}\Omega\bar{z}) = 0$$

which contradicts Lemma 3.

§8. Statement of necessary and sufficient conditions for a Riemann matrix. The principal matrix

We have gradually accumulated facts about a Riemann matrix Ω, starting from the assumption that at least one Abelian function $f(u) = \varphi(u)/\psi(u)$ exists. Now we need to collect these facts into a single statement and show that the necessary conditions are also sufficient. The most convenient way to express these conditions is by using the matrix N^{-1}, the inverse of the matrix N. Since N is nonsingular and skew-symmetric, and its entries are real integers, it follows that N^{-1} is also nonsingular and skew-symmetric and that its entries are rational numbers.

At this point we need to establish several important preliminary relations. Let us set

$$\Delta = \begin{pmatrix} \Omega \\ \Lambda \end{pmatrix} \quad \text{and} \quad J = \begin{pmatrix} 0 & E_p \\ -E_p & 0 \end{pmatrix} . \tag{III.8.1}$$

We know that

$$N = \widetilde{\Omega}\Lambda - \Lambda\widetilde{\Omega} . \tag{III.8.2}$$

We now show that N can also be represented in the form

$$N = \widetilde{\Delta} J \Delta . \tag{III.8.3}$$

Indeed,

$$(\widetilde{\Omega}, \widetilde{\Lambda}) \begin{pmatrix} 0 & E_p \\ -E_p & 0 \end{pmatrix} \begin{pmatrix} \Omega \\ \Lambda \end{pmatrix} = (-\widetilde{\Lambda}, \widetilde{\Omega}) \begin{pmatrix} \Omega \\ \Lambda \end{pmatrix} = -\widetilde{\Lambda}\Omega + \widetilde{\Omega}\Lambda = N .$$

But N is nonsingular since its determinant $|N| = \delta^2 = |\Delta|^2 \neq 0$ by Theorem III.7.3. Hence we deduce from (III.8.3) that

$$\widetilde{\Delta} J \Delta N^{-1} = E_{2p} .$$

If we notice in addition that $J^{-1} = -J$ we get

$$J\Delta N^{-1} = \widetilde{\Delta}^{-1} , \qquad J\Delta N^{-1}\widetilde{\Delta} = E_{2p} , \qquad \Delta N^{-1}\widetilde{\Delta} = -J ,$$

i.e.,

$$\begin{pmatrix}\Omega\\ \Lambda\end{pmatrix} N^{-1}(\tilde{\Omega}, \tilde{\Lambda}) = \begin{pmatrix}\Omega N^{-1}\tilde{\Omega}, & \Omega N^{-1}\tilde{\Lambda}\\ \Lambda N^{-1}\tilde{\Omega}, & \Lambda N^{-1}\tilde{\Lambda}\end{pmatrix} = \begin{pmatrix}0, & -E_p\\ E_p, & 0\end{pmatrix}.$$

Thus we obtain the following relations among the matrices Ω, Λ, and N^{-1}:

$$\Omega N^{-1}\tilde{\Omega} = \Lambda N^{-1}\tilde{\Lambda} = 0, \tag{III.8.4}$$

$$-\Omega N^{-1}\tilde{\Lambda} = \Lambda N^{-1}\tilde{\Omega} = E_p. \tag{III.8.5}$$

We now turn from the Hermitian matrix H to the inverse matrix H^{-1}, which must also be Hermitian just as $\overline{H}$ is (section III.7). Using (III.8.2), (III.7.8), (III.7.23), and (III.7.24) we derive

$$N\overline{F}_1 = (\tilde{\Omega}\Lambda - \tilde{\Lambda}\Omega)\overline{F}_1 = \tilde{\Omega}\Lambda\overline{F}_1 = \tilde{\Omega}\tilde{F}_1 N\overline{F}_1 = \frac{1}{\pi i}\tilde{\Omega}H.$$

Hence $NF_1 = -\tilde{\overline{\Omega}}\overline{H}/\pi i$, from which it follows that $\tilde{F}_1 N = \tilde{\overline{H}}\overline{\Omega}/\pi i = H\overline{\Omega}/\pi i$. Therefore

$$H = \pi i\tilde{F}_1 N\overline{F}_1 = \pi i(\tilde{F}_1 N)N^{-1}(N\overline{F}_1) = \frac{1}{\pi i}(H\overline{\Omega})N^{-1}(\tilde{\Omega}H),$$

or, by multiplying on the left and on the right by H^{-1}, we have

$$H^{-1} = \frac{1}{\pi i}\overline{\Omega}N^{-1}\tilde{\Omega}, \qquad \overline{H}^{-1} = \frac{i}{\pi}\Omega N^{-1}\tilde{\overline{\Omega}}. \tag{III.8.6}$$

This, then, is the required representation of the hermitian matrix $\overline{H}^{-1}$, which is positive together with H.

The facts established in sections III.6 and III.7 allow us to state the next theorem, which summarizes the results of this investigation. The inclusion of the conditions (A), (B), and (C) in the theorem is dictated by the fact that these conditions, which are necessary at this point, will turn out to be sufficient as well.

THEOREM III.8.1. *If $f(u)$ is an Abelian function whose periods are generated by the matrix Ω, then there exists a reduced representation $f(u) = \varphi(u)/\psi(u)$, where $\varphi(u)$ and $\psi(u)$ are Jacobian functions which determine the second period matrix Λ. The characteristic matrix $N = \tilde{\Omega}\Lambda - \tilde{\Lambda}\Omega$, all of whose elements are integers, does not depend on the choice of the functions $\varphi(u)$ and $\psi(u)$; it is skew-symmetric and its determinant is $|N| = \delta^2 \neq 0$. With the aid of the matrix N^{-1}, whose elements are rational numbers, the conditions which the Riemann matrix Ω must necessarily satisfy can be reduced to three major ones:*

(A) $\tilde{N}^{-1} = -N^{-1}$;

(B) $\Omega N^{-1}\tilde{\Omega} = 0$ *is the matrix form of the Riemann equations;*

(C) $i\Omega N^{-1}\tilde{\bar{\Omega}} > 0$ (*i.e., the Hermitian form* $i\tilde{u}\Omega N^{-1}\tilde{\bar{\Omega}}\bar{u}$ *is positive definite*) *is the matrix form of the Riemann inequality.*

It is obvious that the theorem is still true if we substitute a matrix P of the following form for the matrix N^{-1}:

$$P = \lambda N^{-1}, \tag{III.8.7}$$

where λ is a positive integer. In particular, we can choose λ so that all the entries of the matrix (III.8.7) are integers.

In general, we will call a $(2p, 2p)$ matrix P *principal* (with respect to a given $(p, 2p)$ matrix Ω) if its elements are integers and it satisfies the following three conditions:

(A′) $\tilde{P} = -P$, i.e., P is skew-symmetric;

(B′) $\Omega P\tilde{\Omega} = 0$, i.e., P satisfies the Riemann equalities;

(C′) the matrix $i\Omega P\tilde{\bar{\Omega}} > 0$ (or < 0), i.e., P (or $-P$) satisfies the Riemann inequalities.

Then the preceding theorem can be stated in the following more general form.

THEOREM III.8.1′. *For a* $(p, 2p)$ *matrix* Ω *to be a Riemann matrix, it is necessary that there is a matrix* P *that is principal with respect to* Ω.

In the following sections we will show that this condition is also sufficient. The necessity of the condition was discovered in the work of Riemann and Weierstrass; the sufficiency was established by Wirtinger, Picard, and Poincaré. We will prove the following statement.

FUNDAMENTAL THEOREM III.8.2. *For a* $(p, 2p)$ *matrix* Ω *to be a Riemann matrix it is necessary and sufficient that there is a matrix* P *that is principal with respect to* Ω.

We have already established the necessity of the conditions (Theorem III.8.1, where $P = \lambda N^{-1}$).

The proof of the sufficiency involves the following four stages:

(1) Starting from a principal matrix P, a characteristic matrix N defined by the conditions $N^{-1} = \lambda P$ must be constructed, where λ is a rational number which is positive or negative depending on whether $i\Omega P\tilde{\bar{\Omega}} > 0$ or $i\Omega P\tilde{\bar{\Omega}} < 0$.

(2) From the equation $\tilde{\Omega}\Lambda - \tilde{\Lambda}\Omega = N$ all possible second matrices of period Λ must be determined.

(3) We must construct all Jacobian functions corresponding to the matrices Ω and Λ of periods and an arbitrarily chosen parametric vector γ, and express these functions in terms of theta functions of p complex variables.

(4) We must express an Abelian function $f(u)$, for which Ω is the fundamental period matrix, in terms of Jacobian functions of the same type $\{\Omega, \Lambda, \gamma\}$.

We first clarify some simple questions which readers may ask about the conditions (A), (B), and (C) or, more generally, about the conditions (A′), (B′) and (C′).

Why is it not mentioned in condition (A′) that $|P| \neq 0$ (or that $|N^{-1}| \neq 0$ in condition (A))? Why are *the elementary conditions* of section III.1 forgotten?

The answer to these questions is naturally that the requirement $|P| \neq 0$ and "the elementary conditions" are simple consequences of (A′), (B′) and (C′). We denote, for the sake of brevity, the (p, p) matrix $\Omega P \widetilde{\overline{\Omega}}$ by K; then $\widetilde{K} = \overline{\Omega} \widetilde{P} \widetilde{\Omega} = -\overline{\Omega} P \widetilde{\Omega}$. It follows from condition (C′) that $|K| \neq 0$. Otherwise we would also have $|\overline{K}| = 0$ and therefore a nonzero p-vector u would exist, such that $\overline{K} u = 0$, i.e., $K \bar{u} = 0$. Hence $i\tilde{u} K \bar{u} = 0$, whereas by (C′) it must be true that $i\tilde{u} K \bar{u} \neq 0$. Thus $|K| \neq 0$.

For the $(2p, 2p)$ matrix $F = \binom{\Omega}{\overline{\Omega}}$ we derive

$$F P \widetilde{F} = \begin{pmatrix} \Omega \\ \overline{\Omega} \end{pmatrix} P(\widetilde{\Omega}, \widetilde{\overline{\Omega}}) = \begin{pmatrix} \Omega P \widetilde{\Omega}, & \Omega P \widetilde{\overline{\Omega}} \\ \overline{\Omega} P \widetilde{\Omega}, & \overline{\Omega} P \widetilde{\overline{\Omega}} \end{pmatrix}.$$

Because of (B′) two matrices on the "main diagonal" are equal to zero (from $\Omega P \widetilde{\Omega} = 0$ it follows also that $\overline{\Omega} P \widetilde{\overline{\Omega}} = 0$). Hence

$$F P \widetilde{F} = \begin{pmatrix} 0 & K \\ -\widetilde{K} & 0 \end{pmatrix},$$

or, taking determinants, we have

$$|F|\,|P|\,|\widetilde{F}| = |K|^2 > 0;$$

from which it follows that

$$|F| \neq 0, \qquad |P| > 0. \tag{III.8.8}$$

It remains only to make the additional remark that all the "elementary properties" of section III.1. are consequences of the fact that $|F| \neq 0$.

We will show that the conditions (A′), (B′) and (C′) (and (A), (B), (C), in particular) are preserved under admissible transformations of the matrix Ω, i.e., under the transformations

$$\Omega^* = A \Omega M, \tag{III.8.9}$$

where A is a nonsingular (p, p) matrix with complex entries and M is a unimodular $(2p, 2p)$ matrix (with integer entries). We need only transform the principal matrix simultaneously with (III.8.9), replacing it by the matrix

$$P^* = M^{-1} P \widetilde{M}^{-1}. \tag{III.8.10}$$

Recall for comparison (section III.6) that the characteristic matrix N was replaced, under a transformation of the form (III.8.9), by the formula (III.6.12): $N^* = \widetilde{M} N M$, from which it follows that

$$N^{*-1} = M^{-1} N^{-1} \widetilde{M}^{-1}. \tag{III.8.10′}$$

If we simultaneously perform the transformations (III.8.9) and (III.8.10), we obtain

$$\Omega^* P^* \widetilde{\Omega}^* = (A\Omega M)(M^{-1}P\widetilde{M}^{-1})(\widetilde{M}\widetilde{\Omega}\widetilde{A}) = A(\Omega P\widetilde{\Omega})\widetilde{A} = 0,$$

i.e., condition (B′) holds for Ω^* and P^*.

In exactly the same way (taking into consideration that $\overline{M} = M$) we get

$$i\Omega^* P^* \widetilde{\overline{\Omega}}^* = i(A\Omega M)(M^{-1}P\widetilde{M}^{-1})(\widetilde{M}\widetilde{\overline{\Omega}}\widetilde{\overline{A}}) = iA(\Omega P\widetilde{\overline{\Omega}})\widetilde{\overline{A}}.$$

Set $v^* = \widetilde{A}^{-1}u$. Then the conditions $u \neq 0$ and $v \neq 0$ will be equivalent (since $|A| \neq 0$). As a consequence of the fact that

$$i\tilde{v}^*\Omega^* P^* \widetilde{\overline{\Omega}}^* \bar{v}^* = i\tilde{u}A^{-1}(A\Omega P\widetilde{\overline{\Omega}}\widetilde{\overline{A}})\widetilde{\overline{A}}^{-1}u = i\tilde{u}\Omega P\Omega\bar{u},$$

we conclude that the condition (C′) also holds.

We can choose the transformations M and A for the matrices Ω and P so that that the matrices take the simplest form. For normalization purposes we can demand that the greatest common divisor of the elements of P should be equal to 1 (if it is larger than 1, we can divide P by it).

We now use the corollary of the Frobenius theorem (Appendix A.2). By this corollary there is a unimodular $(2p, 2p)$ matrix M such that

$$P^* = MP\widetilde{M} = \begin{pmatrix} 0 & B \\ -B & 0 \end{pmatrix}, \quad \text{where} \quad B = \begin{pmatrix} \beta_1 & 0 & \cdots & 0 \\ 0 & \beta_2 & \cdots & 0 \\ \cdots & \cdots & \cdots & \cdots \\ 0 & 0 & \cdots & \beta_p \end{pmatrix}, \tag{III.8.11}$$

and $\beta_1 = 1, \beta_2, \ldots, \beta_p$ are natural numbers such that $\beta_1 | \beta_2 | \cdots | \beta_{p-1} | \beta_p$. We apply the transformation $\Omega^* = \Omega M^{-1}$ to the first period matrix Ω in accordance with the formulas (III.8.9) and (III.8.10). According to what we have said, P^* will be the principal matrix for Ω^* if and only if P is the principal matrix for Ω. We drop the asterisks in the notation for P and Ω. The advantage of changing the original matrices is to have the principal matrix in canonical form (III.8.11). Let $\Omega = (\Omega_1, \Omega_2)$, where Ω_1 and Ω_2 are (p, p) matrices. Then the matrix in condition (C′) can be written in the form

$$\begin{aligned} i\Omega P\widetilde{\overline{\Omega}} &= i(\Omega_1, \Omega_2)\begin{pmatrix} 0 & B \\ -B & 0 \end{pmatrix}\begin{pmatrix} \widetilde{\overline{\Omega}}_1 \\ \widetilde{\overline{\Omega}}_2 \end{pmatrix} = i(\Omega_1, \Omega_2)\begin{pmatrix} B\widetilde{\overline{\Omega}}_2 \\ -B\widetilde{\overline{\Omega}}_1 \end{pmatrix} \\ &= i(\Omega_1 B\widetilde{\overline{\Omega}}_2 - \Omega_2 B\widetilde{\overline{\Omega}}_1). \end{aligned} \tag{III.8.12}$$

Hence the condition $|\Omega_1| \neq 0$ is necessarily satisfied. Otherwise there is a vector $u \neq 0$ such that $\tilde{u}\Omega_1 = 0$, $\widetilde{\overline{\Omega}}_1\bar{u} = 0$ and therefore

$$i\tilde{u}\Omega P\widetilde{\overline{\Omega}}\bar{u} = i(\tilde{u}\Omega_1 B\widetilde{\overline{\Omega}}_2\bar{u} - \tilde{u}\Omega_2 B\widetilde{\overline{\Omega}}_1\bar{u}) = 0$$

which contradicts (C′).

If B_0 is any nonsingular (p, p) matrix (for example, if $B_0 = E_p$), then by the change of variables $u^* = B_0\Omega_1^{-1}u$ we can transform Ω into the form

$$\Omega^* = B_0\Omega_1^{-1}\Omega = (B_0, C_0). \tag{III.8.13}$$

It follows from (III.8.10) that the principal matrix P^* remains the same. Indeed, the transformation (III.8.13) is the special case of (III.8.9) when $M = E_{2p}$ and $A = B_0\Omega_1^{-1}$.

By admissible linear transformations of the form (III.8.9) and (III.8.10), we reduce the matrices P and Q to the form

$$P = \begin{pmatrix} 0 & B \\ -B & 0 \end{pmatrix}, \qquad B = \begin{pmatrix} \beta_1 & \cdots & 0 \\ \cdots & \cdots & \cdots \\ 0 & \cdots & \beta_p \end{pmatrix}, \tag{III.8.14}$$
$$\beta_1 = 1|\beta_2|\cdots|\beta_{p-1}|\beta_p, \qquad \Omega = (B_0, C_0),$$

where B_0 is any nonsingular (p, p) matrix. In particular, set

$$B_0 = \pi i B^{-1} = \pi i \begin{pmatrix} \beta_1^{-1} & \cdots & 0 \\ \cdots & \cdots & \cdots \\ 0 & \cdots & \beta_p^{-1} \end{pmatrix}.$$

Then the corresponding (p, p) matrix $C_0 = A$ must be symmetric, since it immediately follows from the condition (B$'$) that

$$\Omega P\widetilde{\Omega} = (\pi i B^{-1}, A)\begin{pmatrix} 0 & B \\ -B & 0 \end{pmatrix}\begin{pmatrix} \pi i B^{-1} \\ \widetilde{A} \end{pmatrix}$$
$$= (\pi i B^{-1}, A)\begin{pmatrix} B\widetilde{A} \\ -\pi i E_p \end{pmatrix} = \pi i(\widetilde{A} - A) = 0.$$

We denote the real and imaginary parts of A by A' and A''. The formula (III.8.12) gives us the following formula when $\Omega_1 = \pi i B^{-1}$ and $\Omega_2 = A = A' + iA''$:

$$i\Omega P\widetilde{\overline{\Omega}} = i[\pi i B^{-1}B(A' - iA'') + \pi i(A' + iA'')BB^{-1}] = -2\pi A'. \tag{III.8.15}$$

Therefore the form $\tilde{u}A'\bar{u}$ must be definite (positive or negative). We have therefore established that *one of the possible standard forms for matrices* Ω *and* P *satisfying conditions* (A$'$), (B$'$) *and* (C$'$) *is as follows*:

$$P = \begin{pmatrix} 0 & B \\ -B & 0 \end{pmatrix}, \qquad B = \begin{pmatrix} \beta_1 & \cdots & 0 \\ \cdots & \cdots & \cdots \\ 0 & \cdots & \beta_p \end{pmatrix},$$
$$\beta_1 = 1|\beta_2|\cdots|\beta_{p-1}|\beta_p; \qquad \Omega = (\pi i B^{-1}, A), \tag{III.8.16}$$

and moreover the Hermitian form $\tilde{u}A'\bar{u}$ (*where* $A' = \operatorname{Re} A$) *must be definite.* The equality (III.8.15) shows that the sign of this form is opposite to the sign of the form $\tilde{u}i\Omega P\widetilde{\overline{\Omega}}\bar{u}$.

All this allows us to clarify (in the particular cases $p = 1$ and $p = 2$) the relation between the "elementary" conditions and the conditions expressed by the fundamental theorem.

First let $p = 1$ and let $\Omega = (\omega^{(1)}, \omega^{(2)})$, where $\omega^{(1)}$ and $\omega^{(2)}$ are complex numbers. The "elementary" conditions are reduced to the fact that the determinant $F = \left|\frac{\Omega}{\bar{\Omega}}\right| \neq 0$, i.e., that the numbers $\omega^{(1)}$ and $\omega^{(2)}$ are real independent. Hence by applying the linear transformation $u^* = \pi i u/\omega^{(1)}$, the matrix Ω can be expressed in the equivalent form $\Omega = (\pi i, \tau)$, where $\tau = \tau' + i\tau''$ and moreover $A' = \tau' \neq 0$. It is obvious that the principal matrix here is $P = \left(\begin{smallmatrix} 0 & 1 \\ -1 & 0 \end{smallmatrix}\right)$ (compare (III.8.16)). Thus in the case when $p = 1$ and Abelian functions are elliptic the "elementary" conditions guarantee the existence of the principal matrix.

This phenomenon is no longer true when $p = 2$ (and the more so when $p > 2$). The "elementary" conditions (which again manifest themselves in the fact that $\left|\frac{\Omega}{\bar{\Omega}}\right| \neq 0$, i.e., that the periods $\omega^{(1)}, \omega^{(2)}, \omega^{(3)}, \omega^{(4)}$, given by the columns of a $(2, 4)$ matrix Ω, are real independent) allow us to transform Ω to an equivalent form

$$\Omega = \begin{pmatrix} \pi i \beta_1^{-1} & 0 & s & t \\ 0 & \pi i \beta_2^{-1} & \sigma & \tau \end{pmatrix},$$

where $\beta_1 = 1$ and $\beta_2 = \beta$ are natural numbers and s, t, σ and τ are complex numbers. Setting $s = s' + is''$, $t = t' + it''$, $\sigma = \sigma' + i\sigma''$, $\tau = \tau' + i\tau''$ we obtain

$$|F| = \left|\frac{\Omega}{\bar{\Omega}}\right| = \begin{vmatrix} 0 & 0 & 2s' & 2t' \\ 0 & 0 & 2\sigma' & 2\tau' \\ -\pi i \beta_1^{-1} & 0 & s' - is'' & t' - it'' \\ 0 & -\pi i \beta_2^{-1} & \sigma' - i\sigma'' & \tau' - i\tau'' \end{vmatrix}$$
$$= 4\pi^2 \beta_1^{-1} \beta_2^{-1} (s'\tau' - \sigma' t').$$

Hence an "elementary" condition has the form

$$s'\tau' - \sigma' t' \neq 0. \qquad \text{(III.8.17)}$$

Meanwhile the existence of a principal matrix of the form

$$P = \begin{pmatrix} 0 & 0 & 1 & 0 \\ 0 & 0 & 0 & \beta \\ -1 & 0 & 0 & 0 \\ 0 & -\beta & 0 & 0 \end{pmatrix}$$

requires in this case that the matrix Ω corresponding to it has the special form (III.8.16),

$$\Omega = \begin{pmatrix} \pi i & 0 & s & \sigma \\ 0 & \frac{\pi i}{\beta} & \sigma & \tau \end{pmatrix},$$

i.e., that, first, the following condition of symmetry must hold:

$$t = \sigma;$$

without this (as we have seen) the condition (B′) cannot hold. But the relation (III.8.18) cannot follow from (III.8.17).

Further, if (III.8.18) were true, the condition (C′) would require the definiteness of the matrix

$$A' = \begin{pmatrix} s' & \sigma' \\ \sigma' & \tau' \end{pmatrix},$$

where $s = s' + is''$, $\sigma = \sigma' + i\sigma''$ and $\tau = \tau' + i\tau''$. This requires, as is known from linear algebra, ([20]) that either

$$s' > 0 \quad \text{and} \quad s'\tau' - (\sigma')^2 > 0 \tag{III.8.19}$$

(in the positive definite case) or

$$s' < 0 \quad \text{and} \quad s'\tau' - (\sigma')^2 > 0 \tag{III.8.20}$$

(in the negative definite case). Neither one follows from the elementary condition (III.8.17), which is simply reduced to the following form when $t = \sigma$:

$$s'\tau' - (\sigma')^2 \neq 0.$$

Picard showed, when $p = 2$, that if just the "elementary" conditions hold for the matrix $\Omega(\omega^{(1)}, \dots, \omega^{(4)})$ one still can construct an analytic function $f(u_1, u_2)$ with basic periods $\omega^{(j)}$, $j = 1, 2, 3, 4$. However, this function will not be meromorphic; it will have essential singularities at some finite points of the space $\mathbf{C}^4$.

([20]) By the way, the calculation is so simple that it can be reproduced here. Let a, b, and c be real numbers and let the form

$$F(u, v) = (u, v)\begin{pmatrix} a & b \\ b & c \end{pmatrix}\begin{pmatrix} \bar{u} \\ \bar{v} \end{pmatrix} = au\bar{u} + b(u\bar{v} + v\bar{u}) + cv\bar{v}$$

be positive (the case of a negative form is reduced to this by changing F into $-F$). Then $F(1, 0) = a$, from which it follows that $a > 0$. Set $u = re^{i\alpha}$ and $v = \rho e^{i\beta}$; we obtain

$$F(u, v) = ar^2 + 2br\rho\cos(\alpha + \beta) + c\rho^2 \geq ar^2 - 2|b|r\rho + c\rho^2 \qquad (a > 0).$$

Since there is equality when $\cos(\alpha + \beta) = -1$, a necessary and sufficient condition for the positivity of F is the positivity of a quadratic polynomial $ax^2 - 2|b|x + c$, i.e., the inequality

$$ac - b^2 > 0 \quad (\text{when } a > 0).$$

CHAPTER IV

Construction of Jacobian Functions of a Given Type. Theta Functions and Abelian Functions. Abelian and Picard Manifolds

§1. Construction of the characteristic matrix N and the second period matrix Λ, using given Ω and P

We now begin the proof of the fundamental theorem, in accordance with the plan outlined in section III.8. We start from the fact that Ω and P can be expressed (after appropriate transformations of the type (III.8.9) and (III.8.10)) in the form (III.8.16)

$$P=\begin{pmatrix} 0 & B \\ -B & 0 \end{pmatrix}, \qquad B=\begin{pmatrix} B_1 & 0 & \cdots & 0 \\ \cdots & \cdots & \cdots & \cdots \\ 0 & 0 & \cdots & \beta_p \end{pmatrix}, \qquad 1=\beta_1|\beta_2|\cdots|\beta_p,$$

$$\Omega=(\pi i B^{-1}, A'+iA''), \tag{IV.1.1}$$

where A' and A'' are symmetric (p, p) matrices of real numbers and the matrix A' is definite (positive or negative). In this section we will construct (using P and Ω) the matrices corresponding to P and Ω: the characteristic matrix N and the second period matrix Λ.

We begin with the fact that if N is the characteristic matrix, then its inverse N^{-1} becomes (after multiplication of its entries by an appropriate integer) *the principal matrix* with respect to Ω. Therefore there is a relation

$$nN^{-1}=P \qquad (n\neq 0, \text{ an integer}), \tag{IV.1.2}$$

which can be used to find N and from which it follows that

$$N=nP^{-1}=n\begin{pmatrix} 0 & -B^{-1} \\ B^{-1} & 0 \end{pmatrix}, \tag{IV.1.3}$$

where the integer n can be chosen in such a way that all the elements of N

are integers, i.e., the elements of the diagonal matrix

$$nB^{-1} = \begin{pmatrix} n\beta^{-1} & 0 & \cdots & 0 \\ 0 & n\beta_2^{-1} & \cdots & 0 \\ 0 & 0 & \cdots & n\beta_p^{-1} \end{pmatrix} \tag{IV.1.4}$$

must be integers.

Therefore we need to chose n to be an integral multiple of the number β_p:

$$n = l\beta_p, \qquad l \in Z, \ l \neq 0. \tag{IV.1.5}$$

Since, by Theorem III.8.1, the Hermitian form $i\tilde{u}\Omega N^{-1}\tilde{\overline{\Omega}}\bar{u}$ is positive definite, the number n in (IV.1.2) (or l in (IV.1.5)) must be positive if the form $i\tilde{u}\Omega P\tilde{\overline{\Omega}}$ is positive definite and negative if the form is negative definite.

In order to complete the normalization of the principal matrix we *agree* that the form $i\tilde{u}\Omega P\tilde{\overline{\Omega}}$ is *positive definite.* Then from the relation (III.8.15) we deduce that the form $\tilde{u}A'\bar{u}$ is *negative definite*:

$$\tilde{u}A'\bar{u} < 0, \quad \text{if } u \neq 0. \tag{IV.1.6}$$

In accordance with this, l and n are *positive integers* and in view of (IV.1.5) we find the following representation for nB^{-1}:

$$nB^{-1} = l\begin{pmatrix} \delta_p & 0 & \cdots & 0 \\ 0 & \delta_{p-1} & \cdots & 0 \\ \cdots & \cdots & \cdots & \cdots \\ 0 & 0 & \cdots & \delta_1 \end{pmatrix}, \qquad \delta_j = \frac{\beta_p}{\beta_{p-j+1}}, \qquad j = 1, \dots, p. \tag{IV.1.7}$$

It is obvious that both δ_j and β_k are natural numbers and moreover

$$\delta_1 = 1|\delta_2|\cdots|\delta_p = \beta_p. \tag{IV.1.8}$$

Indeed, $\delta_{j+1}/\delta_j = \beta_{p-j+1}/\beta_{p-j}$.

For the determinant of the characteristic matrix N, we obtain

$$|N| = n^{2p}|B^{-1}|^2 = \frac{n^{2p}}{\beta_1^2\cdots\beta_p^2} = l^{2p}\delta_p^2\cdots\delta_1^2 \tag{IV.1.9}$$

from (IV.1.3), (IV.1.1), and (IV.1.7). Thus its smallest possible value is $\delta_p^2\delta_{p-1}^2\cdots\delta_1^2$ (when $l = 1$). We denote by N_0 the characteristic matrix corresponding to $l = 1$. Then

$$N = lN_0 \qquad (l = 1, 2, \dots). \tag{IV.1.10}$$

We now define the *second period matrix* Λ by using the equation

$$\tilde{\Omega}\Lambda - \tilde{\Lambda}\Omega = N. \tag{IV.1.11}$$

Notice that if Λ_0 is a $(p, 2p)$ matrix which satisfies this equation, then

$$\Lambda = \Lambda_0 + \Sigma\Omega \tag{IV.1.12}$$

(where Σ is any symmetric (p, p) matrix) will also satisfy it.

In fact,

$$\widetilde{\Omega}(\Lambda_0 + \Sigma\Omega) - (\widetilde{\Lambda}_0 + \widetilde{\Omega}\Sigma)\Omega = \widetilde{\Omega}\Lambda_0 - \widetilde{\Lambda}_0\Omega = N.$$

Writing the required matrix Λ in the form

$$\Lambda = (\Lambda_1, \Lambda_2), \tag{IV.1.13}$$

where Λ_1 and Λ_2 are (p, p) matrices, we use the freedom of choice of the matrix Σ in (IV.1.2) to obtain $\Lambda_1 = 0$. Let us show that this is always possible. For this purpose, we substitute (IV.1.13) into (IV.1.11), using the expressions (IV.1.1) for Ω and (IV.1.3) for N. We obtain

$$\begin{aligned}
&\begin{pmatrix} \pi i B^{-1} \\ A \end{pmatrix} (\Lambda_1, \Lambda_2) - \begin{pmatrix} \widetilde{\Lambda}_1 \\ \widetilde{\Lambda}_2 \end{pmatrix} (\pi i B^{-1}, A) \\
&= \begin{pmatrix} \pi i B^{-1}\Lambda_1 & \pi i B^{-1}\Lambda_2 \\ A\Lambda_1 & A\Lambda_2 \end{pmatrix} - \begin{pmatrix} \pi i \widetilde{\Lambda}_1 B^{-1} & \widetilde{\Lambda}_1 A \\ \pi i \widetilde{\Lambda}_2 B^{-1} & \widetilde{\Lambda}_2 A \end{pmatrix} \\
&= \begin{pmatrix} \pi i B^{-1}\Lambda_1 - \pi i \widetilde{\Lambda}_1 B^{-1} & \pi i B^{-1}\Lambda_2 - \widetilde{\Lambda}_1 A \\ A\Lambda_1 - \pi i \widetilde{\Lambda}_2 B^{-1} & A\Lambda_2 - \widetilde{\Lambda}_2 A \end{pmatrix} \\
&= \begin{pmatrix} 0 & -nB^{-1} \\ nB^{-1} & 0 \end{pmatrix}.
\end{aligned} \tag{IV.1.14}$$

From this it follows, in particular, that

$$\pi_i B^{-1}\Lambda_2 - \widetilde{\Lambda}_1 A = -nB^{-1}. \tag{IV.1.15}$$

This means that for a given n (i.e., for a fixed characteristic matrix N) the matrix Λ_2 has a unique expression in terms of Λ_1:

$$\Lambda_2 = -\frac{n}{\pi i} E_p + \frac{1}{\pi i} B\widetilde{\Lambda}_1 A. \tag{IV.1.16}$$

From (IV.1.14) we also deduce for Λ_1 that

$$B^{-1}\Lambda_1 - \widetilde{\Lambda}_1 B^{-1} = 0,$$

i.e., the matrix $B^{-1}\Lambda_1 = \Sigma_1$ is *symmetric*. But then $\Sigma = (\pi i)^{-1} B\Sigma_1 B = (\pi i)^{-1}\Lambda_1 B$ is also symmetric. We then substitute this expression into (IV.1.12), rewritten in the form $\Lambda_0 = \Lambda - \Sigma\Omega$:

$$\Lambda_0 = (\Lambda_1, \Lambda_2) - \Sigma \cdot (\pi i B^{-1}, A) = (\Lambda_1 - \pi i \Sigma B^{-1}, \Lambda_2 - \Sigma A).$$

With the preceding choice for Σ we have

$$\Lambda_1 - \pi i \Sigma B^{-1} = 0;$$

again denoting $\Lambda_2 - \Sigma A$ by Λ_2, we represent the second period matrix in the form $\Lambda_0 = (0, \Lambda_2)$. Formula (IV.1.16) now provides the following formula for Λ_2: $\Lambda_2 = -\frac{n}{\pi i} E_p$, i.e.,

$$\Lambda_0 = \left(0, -\frac{n}{\pi i} E_p\right). \tag{IV.1.17}$$

It is easy to verify, by using (IV.1.14), that Λ_0 indeed satisfies (IV.1.11).

It follows from this analysis that there are no other second period matrices except for those represented by the formula (IV.1.12) (for given Ω and N).

The behavior of Jacobian functions of the same type is determined by the matrices Ω, Λ and the parametric vector γ. Let Σ be an arbitrary symmetric (p, p) matrix and let α and c be p-vectors. We investigate how the type of the Jacobian function $\varphi(u)$ changes under the transformation

$$\varphi^*(u) = \exp[\pi i(\tilde{u}\Sigma u + 2\tilde{\alpha}u)]\varphi(u + c). \tag{IV.1.18}$$

This transformation reduces to a translation of the vector u by c and subsequent multiplication of $\varphi(u + c)$ by an exponential with exponent whose quadratic part is determined by a symmetric matrix Σ and whose linear part is determined by the vector α. We set, for brevity,

$$\tilde{u}\Sigma u + 2\tilde{\alpha}u = \Phi(u).$$

If

$$\varphi(u + \omega^{(k)}) = \exp[2\pi i\varepsilon^{(k)}(\tilde{\Lambda} + \gamma)]\varphi(u) \tag{IV.1.19}$$

(see (III.6.2) and (III.6.3)), we have

$$\begin{aligned} \varphi^*(u + \omega^{(k)}) &= \exp[\pi i\Phi(u + \omega^{(k)})]\varphi(u + c + \omega^{(k)}) \\ &= \exp[\pi i\Phi(u + \omega^{(k)}) + 2\pi i\tilde{\varepsilon}^{(k)}(\tilde{\Lambda}u + \tilde{\Lambda}c + \gamma)]\varphi(u + c) \\ &= \exp\{\pi i[\Phi(u + \omega^{(k)}) - \Phi(u)] \\ &\qquad + 2\pi i\tilde{\varepsilon}^{(k)}(\tilde{\Lambda}u + \tilde{\Lambda}c + \gamma)\}\varphi^*(u). \end{aligned} \tag{IV.1.20}$$

Noticing that $\omega^{(k)} = \Omega\varepsilon^{(k)}$, we obtain

$$\begin{aligned} \Phi(u + \omega^{(k)}) - \Phi(u) &= (\tilde{u} + \tilde{\varepsilon}^{(k)}\tilde{\Omega})\Sigma(u + \Omega\varepsilon^{(k)}) + 2\tilde{\alpha}u + 2\tilde{\alpha}\Omega\varepsilon^{(k)} - \tilde{u}\Sigma u - 2\tilde{\alpha}u \\ &= \tilde{\varepsilon}^{(k)}\tilde{\Omega}\Sigma u + \tilde{u}\Sigma\Omega\varepsilon^{(k)} + \tilde{\varepsilon}^{(k)}\tilde{\Omega}\Sigma\Omega\varepsilon^{(k)} + 2\tilde{\alpha}\Omega\varepsilon^{(k)}. \end{aligned}$$

But $\tilde{u}\Sigma\Omega\varepsilon^{(k)} = \tilde{\varepsilon}^{(k)}\tilde{\Omega}\Sigma u$ since $\tilde{\Sigma} = \Sigma$ and moreover $2\tilde{\alpha}\Omega\varepsilon^{(k)} = 2\tilde{\varepsilon}^{(k)}\tilde{\Omega}\alpha$; hence

$$\Phi(u + \omega^{(k)}) - \Phi(u) = 2\tilde{\varepsilon}^{(k)}\tilde{\Omega}\Sigma u + \zeta_k + 2\tilde{\varepsilon}^{(k)}\tilde{\Omega}a, \tag{IV.1.21}$$

where $\zeta_k = \varepsilon^{(k)}\tilde{\Omega}\Sigma\Omega\varepsilon^{(k)}$ is the kth diagonal entry of a symmetric $(2p, 2p)$ matrix $\tilde{\Omega}\Sigma\Omega$. We denote a $2p$-vector with components ζ_k by ζ; then $\zeta_k = \varepsilon^{(k)}\zeta$.

Substituting (IV.1.21) into (IV.1.20), we obtain

$$\begin{aligned} \varphi^*(u + \omega^{(k)}) = \exp\Big\{2\pi i\tilde{\varepsilon}^{(k)}\Big[&(\tilde{\Lambda} + \tilde{\Omega}\Sigma)u \\ &+ \frac{1}{2}\zeta + \tilde{\Omega}\alpha + \tilde{\Lambda}c + \gamma\Big]\Big\}\varphi^*(u). \end{aligned} \tag{IV.1.22}$$

Comparing this with (IV.1.19), we find that the function (IV.1.18) belongs to the type for which the second matrix Λ is replaced by $\Lambda^* = \Lambda + \Sigma\Omega$ and the parametric vector γ is replaced by

$$\gamma^* = \gamma + \frac{1}{2}\zeta + \widetilde{\Omega}\alpha + \widetilde{\Lambda}c; \qquad \text{(IV.1.22}')$$

we recall that ζ is the vector formed by the diagonal elements of the matrix $\widetilde{\Omega}\Sigma\Omega$.

Since the matrix $(\widetilde{\Omega}, \widetilde{\Lambda})$ is nonsingular, i.e., since

$$|\widetilde{\Omega}, \widetilde{\Lambda}| = \begin{vmatrix} \Omega \\ \Lambda \end{vmatrix} = \delta \neq 0$$

(Theorem III.7.3), *for any* $2p$*-vector* γ^* *we can find from the equation* (IV.1.22′) p*-vectors* α *and* c *such that the equation* (IV.1.22) *is satisifed.*

We have therefore proved that *if there exist Jacobian functions* $\varphi(u)$ *of the type determined by the matrices* Ω *and* Λ *and the parametric vector* γ, *then there must be Jacobian functions of the type determined by the matrices* Ω, $\Lambda^* = \Lambda + \Sigma\Omega$ (*where* Σ *is any symmetric* (p, p) *matrix*) *and by a parametric* $2p$*-vector* γ^* *with any components* $\gamma_1^*, \dots, \gamma_{2p}^*$.

§2. Construction of Jacobian functions of a given type. Theta functions

The standard form for the matrices Ω, N, and Λ which we derived in the previous section, and the notation connected with it, are dictated by our desire to follow the tradition dating from the time of Riemann and Weierstrass and solidified under the leadership of Krazer in his book on the theory of theta functions of several variables. ([21])

We start from the following expressions for the basic matrices:

$$\Omega = (\pi i B^{-1}, A), \qquad N = n\begin{pmatrix} 0 & B^{-1} \\ B^{-1} & 0 \end{pmatrix}, \qquad \Lambda = \left(0, -\frac{n}{\pi i}E_p\right). \qquad \text{(IV.2.1)}$$

Here A is a symmetric (p, p) matrix satisfying the condition $\operatorname{Re} A = A' < 0$ and

$$B = \begin{pmatrix} \beta_1 & 0 & \cdots & 0 \\ 0 & \beta_2 & \cdots & 0 \\ \cdots & \cdots & \cdots & \cdots \\ 0 & 0 & \cdots & \beta_p \end{pmatrix}, \qquad nB^{-1} = l\begin{pmatrix} \delta_p & 0 & \cdots & 0 \\ 0 & \delta_{p-1} & \cdots & 0 \\ \cdots & \cdots & \cdots & \cdots \\ 0 & 0 & \cdots & \delta_1 \end{pmatrix}, \qquad \text{(IV.2.2)}$$

and the natural numbers β_j, δ_k, l, and n satisfy the relations

$$\beta_1 = 1|\beta_2|\cdots|\beta_p; \qquad \beta_j = \frac{\delta_p}{\delta_{p-j+1}}, \qquad \delta_k = \frac{\beta_p}{\beta_{p-k+1}}, \qquad j, k = 1, \dots, p;$$
$$\delta_1 = 1|\delta_2|\cdots|\delta_p = \beta_p, \qquad n = l\beta_p = l\delta_p, \qquad l = 1, 2, \dots. \qquad \text{(IV.2.3)}$$

([21]) A. Krazer, *Lehrbuch der Thetafunktionen*, Leipzig, 1903.

To construct Jacobian functions we need to choose the parametric $2p$-vector γ. As we pointed out at the end of section IV.1, the vector γ can have any value. We choose γ in such a way that the equations which characterize the Jacobian functions of the type $\{\Omega, \Lambda, \gamma\}$ are in the simplest possible form. For convenience, we express γ in the form $\gamma = \binom{\xi}{\eta}$ where ξ is the p-vector formed by the first p coordinates of γ and η is the vector formed by the last p coordinates.

We note that in our case the periods $\omega^{(h)}$, $h = 1, \dots, 2p$, separate naturally into two groups:

$$\omega^{(h)} = \pi i B^{-1} e^{(h)} = \frac{\pi i}{\beta_h} e^{(h)}, \qquad h = 1, \dots, p;$$

$$\omega^{(p+k)} = Ae^{(k)}, \qquad k = 1, \dots, p.$$

Here $e^{(h)}$, $e^{(k)}$ are columns of the unit matrix E_p.

Since the first p columns of the matrix Λ consist of zeros, we find from the general formulas characterizing the Jacobian functions, i.e., from

$$\varphi(u + \omega^{(h)}) = \exp[2\pi i \tilde{\varepsilon}^{(h)}(\tilde{\Lambda} + \gamma)]\varphi(u), \qquad h = 1, \dots, 2p \tag{2.4}$$

(recall that $\varepsilon^{(h)}$ is a column of the matrix E_{2p}) that for the periods of the first group:

$$\varphi\left(u + \frac{\pi i}{\beta_h} e^{(h)}\right) = \exp(2\pi i \xi_h)\varphi(u), \qquad h = 1, \dots, p.$$

It is now natural to set $\xi_1 = \cdots = \xi_p = 0$. Then $\varphi(u)$ must satisfy the following conditions of periodicity:

$$\varphi\left(u_1, \dots, u_h + \frac{\pi i}{\beta_h}, \dots, u_p\right) = \varphi_1(u_1, \dots, u_h, \dots, u_p), \qquad h = 1, \dots, p. \tag{IV.2.5}$$

For the periods $Ae^{(k)}$ $(k = 1, \dots, p)$ of the second group, the behavior of the Jacobian matrix $\varphi(u)$ is determined by the matrix $-nE_p/\pi i$ that makes up the second half of the $(p, 2p)$ matrix Λ. Hence

$$\begin{aligned}\varphi(u + Ae^{(k)}) &= \exp\left[2\pi i \tilde{e}^{(k)}\left(-\frac{n}{\pi i}u + \eta\right)\right]\varphi(u) \\ &= \exp(-2nu_k + 2\pi i \eta_k)\varphi(u), \qquad k = 1, \dots, p.\end{aligned}$$

To facilitate further calculations we now set

$$\eta_k = -\frac{n}{2\pi i}\alpha_{kk}, \qquad (k = 1, \dots, p), \tag{IV.2.6}$$

where $\alpha_{11}, \dots, \alpha_{pp}$ are the diagonal entries of the matrix A. Then we have

$$\varphi(u + Ae^{(k)}) = \exp\left[-2n\left(u_k + \frac{1}{2}\alpha_{kk}\right)\right]\varphi(u), \qquad k = 1, \dots, p. \tag{IV.2.7}$$

It is useful to rewrite (IV.2.5) and (IV.2.7), using matrix notation in a more consistent way :

$$\varphi(u+\pi i B^{-1} e^{(h)}) = \varphi(u), \qquad h = 1, \ldots, p; \tag{IV.2.5$'$}$$

$$\varphi(u+Ae^{(k)}) = \exp\left[-2n\tilde{e}^{(k)}\left(u+\frac{1}{2}Ae^{(k)}\right)\right]\varphi(u), \qquad k=1,\ldots,p. \tag{IV.2.7$'$}$$

The problem is to construct all possible entire functions $\varphi(u)$ satisfying the equations (IV.2.5$'$) and (IV.2.7$'$).

It is natural to start from the equations (IV.2.5$'$). These equations imply that the required functions must be representable as Fourier series of the following type (cf. §II.7):

$$\varphi(u) = \sum C_\nu \exp(2\tilde{\nu} B u), \tag{IV.2.8}$$

where the summation is over all p-tuples ν containing p integers $\nu_1, \ldots, \nu_p$ and considered as column matrices (vectors).

Using the explicit coordinates, we can write the exponent $2\tilde{\nu}Bu$ in the following form:

$$2\tilde{\nu}Bu = 2(\nu_1\beta_1 u_1 + \cdots + \nu_p \beta_p u_p). \tag{IV.2.9}$$

The substitution of $u_h + \pi i/\beta_h$ for u_h changes the exponent by $2\nu_h \pi i$ and therefore the value of each term in the series stays the same. It remains to restrict the Fourier coefficients C_ν in such a way that any entire function represented by the series (IV.2.8) also satisfies the equations (IV.2.7$'$). If we can show that a series with these coefficients still converges absolutely and uniformly over any p-disk $|u_h| \le R < +\infty$, $h = 1, \ldots, p$, then the construction of the Jacobian functions of a given type $\{\Omega, \Lambda, \gamma\}$ will be complete.

Substituting the expansion (IV.2.8) into each of the equations (IV.2.7$'$) we obtain

$$\sum_{(\nu)} C_\nu \exp(2\tilde{\nu}Bu) e^{2\tilde{\nu}BAe^{(k)}} = \sum_{(\nu)} C_\nu \exp[2(\nu\widetilde{B} - n\tilde{e}^{(k)})u] e^{-n\tilde{e}^{(k)}Ae^{(k)}},$$
$$k = 1, \ldots, p \tag{IV.2.10}$$

Here we intentionally use two notations for exponential functions in order to separate the part of the exponent that depends on u. In order to have the right to refer to the uniqueness of the Fourier series expansion we must express the linear function $2(\nu B - n\tilde{e}^{(k)})u$ in the form $2\tilde{\mu}Bu$, where μ is a p-vector with integer coordinates. It is obvious that it is connected with ν by the relation

$$\tilde{\nu}B - n\tilde{e}^{(k)} = \tilde{\mu}B, \tag{IV.2.11}$$

from which it follows that

$$\mu = \nu - nB^{-1}e^{(k)}, \qquad \nu = \mu + nB^{-1}e^{(k)}. \tag{IV.2.12}$$

Since the elements of the matrix nB^{-1} are integers (see (IV.2.2)), the inverse transformation (IV.2.11) (or (IV.2.12)) is a one-to-one map of the set of p-vectors with integer coordinates onto itself. Hence μ can serve as a summation index just as well as ν. Assuming that our Fourier expansion converges absolutely (we will prove this later), and replacing ν by μ, we obtain the following formula from the equations (IV.2.10)–(IV.2.12):

$$\sum C_\mu \exp(2\tilde{\mu}Bu)e^{2\tilde{\mu}BAe^{(k)}} = \sum C_{\mu+nB^{-1}e^{(k)}} \exp(2\tilde{\mu}Bu)e^{-n\tilde{e}^{(k)}Ae^{(k)}},$$

and therefore

$$C_{\mu+nB^{-1}e^{(k)}} = C_\mu \exp(2\tilde{\mu}BAe^{(k)} + n\tilde{e}^{(k)}Ae^{(k)}), \qquad k = 1, \dots, p, \quad \text{(IV.2.13)}$$

for any μ.

If we consider the class of all coefficients $\{C_\mu\}$ as a complex valued function defined on the integral lattice formed by the μ's, the equation (IV.2.13) exhibits a deep analogy between C_μ and a Jacobian function. Like a Jacobian function, the coefficients C_μ gain an exponential factor with an exponent of the first degree with respect to μ : $l_k(\mu) + b_k$ if we change μ by the period $nB^{-1}e^{(k)} = (n/\beta_k)e^{(k)}$. It is easy to find a quadratic polynomial $\Phi(\mu)$ such that its differences $\Phi(\mu + nB^{-1}e^{(k)}) - \Phi(\mu)$ coincide with $l_k(\mu) + b_k$, $k = 1, \dots, p$. If we then multiply C_μ by $\exp[\Phi(\mu)]$, we obtain a genuine periodic function of μ. We obtain a simplification in comparison with the discussion of section III.7, where we obtained *only the periodicity of the absolute value* of the product of a Jacobian function $\varphi(u)$ and a similar factor $\exp[\Phi(u)]$, in that analyticity is required of the Jacobian functions whereas this requirement is not needed here. In the case at hand, the function

$$\Phi(\mu) = \frac{1}{n}\tilde{\mu}BAB\mu \qquad \text{(IV.2.14)}$$

satisfies exactly the conditions

$$\begin{aligned}
&\Phi(\mu + nB^{-1}e^{(k)}) - \Phi(\mu) \\
&= \frac{1}{n}(\tilde{\mu} + n\tilde{e}^{(k)}B^{-1})BAB(\mu + nB^{-1}e^{(k)}) - \frac{1}{n}\tilde{\mu}BAB\mu \\
&= 2\tilde{\mu}BAe^{(k)} + n\tilde{e}^{(k)}Ae^{(k)}, \qquad k = 1, \dots, p \qquad \text{(IV.2.15)}
\end{aligned}$$

(one can easily arrive at the expression (IV.2.14) provided one sets $\Phi(\mu) = \tilde{\mu}S\mu + 2T\mu$ where S is a symmetric (p, p) matrix and then restricts $\Phi(\mu)$ to satisfy the conditions (IV.2.15)).

We note that at this point the appropriateness of the choice of the parametric vector $\gamma = \binom{\xi}{\eta}$ is justified; its components appear naturally when the differences of the function $\Phi(\mu) = \tilde{\mu}BAB\mu/n$ are computed.

Now the equations (IV.2.13) take the form

$$C_{\mu+nB^{-1}e^{(k)}} = \exp[\Phi(\mu + nB^{-1}e^{(k)}) - \Phi(\mu)]C_\mu, \qquad k = 1, \dots, p,$$

and this means that $C_\mu \exp[-\Phi(\mu)]$ is *a periodic function* of μ with periods $nB^{-1}e^{(k)}$, $k = 1, \dots, p$. Thus, if we set

$$C_\mu^* = C_\mu \exp[-\Phi(\mu)], \qquad \text{(IV.2.16)}$$

we have

$$C_{\mu+nB^{-1}e^{(k)}}^* = C_\mu^*, \qquad k = 1, \dots, p. \qquad \text{(IV.2.17)}$$

Now consider just p-vectors g with integer coordinates $g_1, \dots, g_p$ which belong to a parallelotope Π constructed from the integral periods

$$\frac{n}{\beta_1}e^{(1)}, \dots, \frac{n}{\beta_p}e^{(p)}.$$

Since $n/\beta_j = l\beta_p/\beta_j = l\delta_{p-j+1}$, Π *contains only a finite number of the corresponding points*, and moreover this number, by (IV.1.9), is

$$l^p\delta_1 \cdots \delta_p = \sqrt{|N|} = |\delta|, \qquad \text{(IV.2.18)}$$

where $|\delta|$ is the "determinant" of the Jacobian function $\varphi(u)$ (see §III.6).

It is obvious that for each point $\mu(\mu_1, \dots, \mu_p)$ with integer coordinates there is one and only one congruent point $g(g_1, \dots, g_p)$. Moreover, g_i is the remainder in the division of μ_j by $l\delta_{p-j+1}$:

$$\mu_j = l\delta_{p-j+1}m_j + g_j = \frac{n}{\beta_j}m_j + g_j, \qquad \text{(IV.2.19)}$$

where m_j is the quotient and $0 \le g_j < n/\beta_j$ $(j = 1, \dots, p)$.

Indeed, if we denote p-vectors with coordinates μ_j, m_j and g_j by μ, m and g, respectively, by (IV.2.19) we have a relation

$$\mu = g + nB^{-1}m, \qquad g \in \Pi, \qquad \text{(IV.2.20)}$$

where $nB^{-1}m$ is a period of the function C_μ^*:

$$nB^{-1}m = \begin{pmatrix} m_1\dfrac{n}{\beta_1} \\ \vdots \\ m_p\dfrac{n}{\beta_p} \end{pmatrix}.$$

Let us denote by $c_\mu[g]$ the periodic function of index μ defined, for each fixed point $g \in \Pi$, by the equalities

$$\begin{gathered} c_g[g] = 1; \qquad c_\mu[g] = 0 \ \text{ if } \mu \in \Pi \ \text{ and } \ \mu \ne g; \\ c_{\mu+nB^{-1}e^{(k)}}[g] = c_\mu[g], \qquad k = 1, \dots, p. \end{gathered} \qquad \text{(IV.2.21)}$$

Then the sum

$$\sum_{g\in\Pi} C_g^* c_\mu[g],$$

over all the points of Π will coincide, for each value of $\mu \in \Pi$, with the corresponding value of C_μ^*. Because of the periodicity of C_μ^* and $c_\mu[g]$, it follows that for *any* μ

$$C_\mu^* = \sum_{g \in \Pi} C_g^* c_\mu[g]. \qquad \text{(IV.2.22)}$$

Using this and (IV.2.16), we find that the Jacobian function (IV.2.8) satisfies

$$\begin{aligned} \varphi(u) &= \sum_\mu C_\mu^* \exp[2\tilde{\mu}Bu + \Phi(\mu)] \\ &= \sum_\mu \sum_{g \in \Pi} C_g^* c_\mu[g] \exp[2\tilde{\mu}Bu + \Phi(\mu)] \\ &= \sum_{g \in \Pi} C_g^* \left(\sum_\mu c_\mu[g] \exp[2\tilde{\mu}Bu + \Phi(\mu)] \right). \end{aligned} \qquad \text{(IV.2.23)}$$

Therefore, *each Jacobian function of a given type* $\{\Omega, \Lambda, \gamma\}$ *is a finite linear combination of Jacobian functions of a special form, and the number of these functions is equal to their "determinant"* $|\delta| = \sqrt{|N|}$.

These functions are called *theta functions of order* n and are denoted by

$$\Theta_n[g](u) = \sum_\mu [g] \exp[2\tilde{\mu}Bu + \Phi(\mu)]. \qquad \text{(IV.2.24)}$$

Thus we arrive at the conclusion that *a theta function* $\Theta_n[g](u)$ *of order* n is defined to be an entire function of p complex variables determined by the Fourier series (IV.2.24).

Here the matrix B and the quadratic form $\Phi(\mu)$ are given by (IV.2.2) and (IV.2.14), and moreover A is a symmetric (p, p) matrix which satisfies the condition $A' = \operatorname{Re} A < 0$. The coefficients $c_\mu[g]$ are defined by the equalities (IV.2.21) for any p-vector μ with integral coordinates.

Using theta functions of order n, *the general Jacobian function* $\varphi(u)$ *of type* $\{\Omega, \Lambda, \gamma\}$ *is expressed* (as was proved just above) *by the formula*

$$\varphi(u) = \sum_{g \in \Pi} C_g^* \Theta_n[g](u), \qquad \text{(IV.2.23}')$$

where the $\sqrt{|N|}$ *coefficients* C_g^* *can take any complex values.*

To complete the construction of the Jacobian functions, it remains to show that the theta series (IV.2.24) converge absolutely and uniformly in each hyperball $\|u\| = \sqrt{|u_1|^2 + \cdots + |u_p|^2} \le R < +\infty$.

In this proof we will make extensive use of the Riemann inequality

$$\tilde{u}A'\bar{u} < 0 \quad \text{if } u \neq 0. \qquad \text{(IV.2.25)}$$

By (IV.2.21), the absolute value of the general term of the series (IV.2.24) can be estimated as follows:

$$|c_\mu[g] \exp[2\tilde{\mu}Bu + \Phi(\mu)]| \le \exp[2\|\tilde{\mu}Bu\| + \operatorname{Re}\Phi(\mu)]. \qquad \text{(IV.2.26)}$$

We have the following estimate for $||\tilde{\mu}Bu||$ in the hyperball $||u|| \le R$:

$$\begin{aligned} \|\tilde{\mu}Bu\| &= |\beta_1\mu_1u_1 + \cdots + \beta_p\mu_pu_p| \le \beta_p\|\mu\|\,\|u\| \\ &\le R\beta_p\|\mu\| . \end{aligned} \tag{IV.2.27}$$

Furthermore, from (IV.2.14) we deduce that

$$\operatorname{Re}\Phi(\mu) = \operatorname{Re}\left[\frac{1}{n}\tilde{\mu}BAB\mu\right] = \frac{1}{n}\tilde{\mu}BA'B\mu .$$

Here we set $B\mu = \nu$, so that

$$\|\nu\| = \sqrt{(\beta_1\mu_1)^2 + \cdots + (\beta_p\mu_p)^2} \ge \|\mu\| ;$$

then $\operatorname{Re}[\Phi(\mu)] = \tilde{\nu}A'\nu/n$ and we can use (IV.2.25) in order to estimate $\operatorname{Re}[\Phi(\mu)]$. If $-\alpha^2 < 0$ is the maximum of the Hermitian form $\tilde{u}A'\bar{u}$ when $||u|| = 1$, for any $u \ne 0$ we have $\tilde{u}A'\bar{u} \le -\alpha^2||u||^2$, from which it follows that

$$\operatorname{Re}\Phi(\mu) = \frac{1}{n}\tilde{\nu}A'\nu \le -\frac{\alpha^2}{n}\|\nu^2\| \le -\frac{\alpha^2}{n}\|\mu\|^2 . \tag{IV.2.28}$$

If we take acount of (IV.2.27) and (IV.2.28), the estimate (IV.2.26) takes the form

$$|c_\mu[g]\exp[2\tilde{\mu}Bu + \Phi(\mu)]| \le \exp\left[R\beta_p\|\mu\| - \frac{\alpha^2}{n}\|\mu\|^2\right] . \tag{IV.2.29}$$

We will apply this to each point μ on the boundary of the parallelotope $P_q : |\mu_j| \le q$, $j = 1, \dots, p$; $q = 1, 2, \dots$. Since P_q contains only $(2q+1)^p$ points with integer coordinates, the number of points on the boundary of P_q does not exceed $(2q + 1)^p$. Moreover, it is obvious that $q \le ||\mu|| \le q\sqrt{p}$ for the points on the boundary of P_q. Hence if we sum the inequalities (IV.2.29) over all these points we get

$$\begin{aligned} S_q &= \sum |c_\mu[g]\exp[2\tilde{\mu}Bu + \Phi(u)]| \\ &\le (2q+1)^q \exp\left[R\beta_p\sqrt{p}q - \frac{\alpha^2}{n}q^2\right] . \end{aligned} \tag{IV.2.30}$$

Here the right-hand side tends to 0 as $q \to \infty$, and this remains true if both sides of the inequality are multiplied by q^2. Hence $S_q < C(R)/q^2$; from this there follows *the absolute and uniform convergence of the theta series and therefore the convergence of the Fourier expansion of any Jacobian function.*

Let us notice some simple properties of theta functions. In the first place, *the* $\sqrt{|N|} = |\delta|$ *theta functions of any order* $n = l\beta_p$ *are linearly independent.*

In fact, assume that for some choice of constants C_g we have

$$\sum_{g\in\Pi} C_g\Theta_n[g](u) = 0 .$$

If we replace $\Theta_n[g](u)$ by the formula (IV.2.24), we obtain a Fourier series (because of the convergence properties proved above) whose sum is zero, i.e.,

$$\sum_{\mu}\left(\sum_{g\in\Pi} C_g c_\mu[g]\exp\Phi(\mu)\right)\exp 2\tilde{\mu}Bu = 0.$$

Hence for each μ

$$\sum_{g\in\Pi} C_g c_\mu[g]\exp\Phi(\mu) = 0.$$

When $\mu = g \in \Pi$ we find that $C_g c_g[g] = C_g = 0$. Since this is true for each $g \in \Pi$, the linear independence of the functions $\Theta_n[g](u)$ of a given order n is established.

Even though the formula (IV.2.24) is so simple, it has the defect that it *does not explicitly separate* out the terms *with coefficients which are definitely not equal to zero.* From the definition of $c_\mu[g]$ it follows that $c_\mu[g] \neq 0$ (and is equal to 1) if and only if $\mu = g + nB^{-1}m$, where m is a p-vector with arbitrary integral coordinates. Retaining only these terms in the series (IV.2.24), we obtain

$$\Theta_n[g](u) = \sum_m \exp[2(\tilde{g} + n\tilde{m}B^{-1})Bu + \Phi(g + nB^{-1}m)], \qquad \text{(IV.2.24}'\text{)}$$

where the summation is over all p-vectors m with integral coordinates.

It is useful to notice that *this formula makes sense for arbitrary vectors* g *even if they do not belong to* Π.

Indeed, if we replace g by a vector $g + nB^{-1}\mu$, where μ is any p-vector with integral coefficients, we obtain

$$\Theta_n[g + nB^{-1}\mu](u) = \sum_m \exp\{[g + n(\tilde{m} + \tilde{\mu})B^{-1}]Bu + \Phi[g + nB^{-1}(m + \mu)]\}.$$

If we denote the variable vector $\mu + m$ again by m, we obtain

$$\Theta_n[g + nB^{-1}\mu](u) = \sum_m \exp[2(\tilde{g} + n\tilde{m}B^{-1})Bu + \Phi(g + nB^{-1}m)] = \Theta_n[g](u). \qquad \text{(IV.2.31)}$$

We now return to (IV.2.24). If we use (IV.2.14), we obtain

$$\begin{aligned}\Phi(g + nB^{-1}m) &= \frac{1}{n}(\tilde{g} + n\tilde{m}B^{-1})BAB(g + nB^{-1}m)\\ &= n\left(\tilde{m} + \frac{1}{n}\tilde{g}B\right)A\left(m + \frac{1}{n}Bg\right).\end{aligned}$$

Similarly, the part of the exponent which is linear in u can be written in the form

$$2(\tilde{g} + n\tilde{m}B^{-1})Bu = 2\left(\tilde{m} + \frac{1}{n}\tilde{g}B\right)nu.$$

Substituting this into (IV.2.24′), we obtain *the standard expression for a theta function of order* n:

$$\Theta_n[g](u) = \sum_m \exp\left\{\left(\tilde{m} + \frac{1}{n}\tilde{g}B\right) n \left[A\left(m + \frac{1}{n}Bg\right) + 2u\right]\right. . \qquad \text{(IV.2.24'')}$$

§3. Construction of Abelian functions. Fields of Abelian functions

We are now able to complete the proof of the fundamental theorem stated in section III.8.

In order for a $(p, 2p)$ matrix Ω to be a Riemann matrix, i.e., in order for there to exist a nondegenerate meromorphic function $f(u)$ of p complex variables $u_1, \ldots, u_p$, such that the set of periods of $f(u)$ coincides exactly with the set generated by the matrix Ω, it is necessary and sufficient that there is at least one matrix P which is principal with respect to Ω.

The necessity of this condition was established in the same section III.8. Assuming that it holds, we have also shown in section IV.1 that *without loss of generality one can construct Jacobian functions* with respect to the equivalence class represented by the matrix Ω (cf. section III.1) *starting with the standard form of the first and second matrices of periods* Ω *and* Λ *and from the parametric vector* γ. Such a construction was made in section IV.2. It now remains (using the same standard form) *to prove the existence of the corresponding Abelian function* $f(u)$.

A Jacobian function $\varphi(u)$ of type $\{\Omega, \Lambda, \gamma\}$ is chosen as the main element of this construction. It is an entire function of p complex variables which satisfies the relations

$$\varphi(u + \Omega\varepsilon^{(k)}) = \exp[2\pi i\tilde{\varepsilon}(\tilde{\Lambda}u + \gamma)]\varphi(u), \qquad \text{(IV.3.1)}$$

or, in more detail (see (IV.2.5′) and (IV.2.7′)),

$$\varphi(u + \pi i B^{-1}e^{(h)}) = \varphi(u), \qquad h = 1, \ldots, p,$$

$$\varphi(u + Ae^{(k)}) = \exp\left[-2n\tilde{e}^{(k)}\left(u + \frac{1}{2}Ae^{(k)}\right)\right]\varphi(u), \qquad k = 1, \ldots, p. \qquad \text{(IV.3.1')}$$

We naturally are assuming that $\varphi(u) \not\equiv 0$.

It was shown in section IV.2 that $\varphi(u)$ can be represented as a linear combination of the corresponding theta functions

$$\varphi(u) = \sum_{g\in\Pi} C_g\Theta_n[g](u), \qquad \text{(IV.3.2)}$$

where not all C_g are equal to 0 (Π is a parallelotope, formed from the p vectors $(n/\beta_1)e^{(1)}, \ldots, (n/\beta_p)e^{(p)}$).

By Theorem B.2.2 (see the Appendix) we can find a point $\tilde{c} = (c_1, \ldots, c_p) \in \mathbf{C}^p$ such that the function $\varphi(u + c)$ will satisfy, for any $u \in \mathbf{C}^p$, the condition

$$(\varphi(u), \varphi(u+c)) = 1. \tag{IV.3.4}$$

This means that no function which is analytic at a point $u \in \mathbf{C}^p$ and zero at that point can be a common divisor of $\varphi(u)$ and $\varphi(u+c)$ at that point. If we substitute $u-c$ for u in (IV.3.3), we obtain

$$(\varphi(u-c), \varphi(u)) = 1. \tag{IV.3.4}$$

It is important to note that the functions $\varphi(u \pm c)$ are also Jacobian. This fact was mentioned earlier (section III.4), but of course it follows immediately from (IV.3.1$'$). For these functions, the formulas take the form

$$\begin{gathered} \varphi(u \pm c + \pi i B^{-1} e^{(h)}) = \varphi(u \pm c), \\ \varphi(u \pm c + Ae^{(k)}) = \exp\left[-2n\tilde{e}^{(k)}\left(u + \frac{1}{2}Ae^{(k)} \pm c\right)\right] \varphi(u \pm c). \end{gathered} \tag{IV.3.1$''$}$$

Consider two meromorphic functions

$$f_1(u) = \frac{\varphi(u+c)}{\varphi(u)}, \qquad f_2(u) = \frac{\varphi(u-c)}{\varphi(u)}. \tag{IV.3.5}$$

Because of (IV.3.3) and (IV.3.4) both representations are reduced, i.e., the numerator and the denominator of each of these fractions do not have any common divisors (taking the value 0 at a given point) for any u. The formulas (IV.3.1$'$) and (IV.3.1$''$) provide the following two pairs of relations for these two functions:

$$\begin{aligned} f_1(u + \pi i B^{-1} e^{(h)}) &= f_1(u), \qquad f_1(u + Ae^{(k)}) = \exp(-2n\tilde{e}^{(k)}c) f_1(u), \\ f_2(u + \pi i B^{-1} e^{(h)}) &= f_2(u), \qquad f_2(u + Ae^{(k)}) = \exp(2n\tilde{e}^{(k)}c) f_2(u). \end{aligned} \tag{IV.3.6}$$

This suggests taking the product of these functions

$$f(u) = f_1(u) f_2(u) = \frac{\varphi(u+c)\varphi(u-c)}{\varphi^2(u)}. \tag{IV.3.7}$$

This meromorphic function is also represented in an irreducible form. Indeed, any prime divisor of the function $\varphi^2(u)$ at some point must be a divisor of the function $\varphi(u)$ at the same point. If it divides the numerator of the fraction (IV.3.6) as well, then by a lemma of Euclid (see Appendix, section B.1) it must divide either $\varphi(u+c)$ or $\varphi(u-c)$. But both of these are impossible because of (IV.3.3) and (IV.3.4).

At the same time, the relations (IV.3.6) yield the following relations for the function (IV.3.7):

$$f(u + \pi i B^{-1} e^{(h)}) = f(u), \qquad f(u + Ae^{(k)}) = f(u), \qquad h, k = 1, \dots, p. \tag{IV.3.8}$$

This means that the Jacobian functions in the numerator and the denominator of the fraction (IV.3.7) belong to the same type.

Thus we have constructed *a meromorphic function* (IV.3.7) *for which each p-vector*

$$\omega = m_1 \pi i B^{-1} e^{(1)} + \cdots + m_p \pi i B^{-1} e^{(p)} + n_1 Ae^{(1)} + \cdots + n_p Ae^{(p)} \tag{IV.3.9}$$

(m_j, n_k are arbitrary integers) *generated by the matrix* Ω *is a period.* We have to prove that *no other p-vector* τ *can be a period of this function.* From this, by the way, it will follow that $f(u)$ does not have infinitesimal periods, i.e., that it is nondegenerate.

It is convenient to break the proof into two parts. First we prove that for each period τ one can find ω (IV.3.9) such that

$$n\tau = \omega \qquad (n = l\beta_p = l\delta_p). \tag{IV.3.10}$$

Then, using the representation (IV.3.2) and putting some additional constraints on $\varphi(u)$, we prove that we can replace n by 1 (for the corresponding ω) in the preceding relation.

Now let

$$f(u+\tau) = f(u) \tag{IV.3.11}$$

for all $u \in \mathbf{C}^p$. Then (IV.3.7) gives

$$\varphi^2(u)\varphi(u+\tau+c)\varphi(u+\tau-c) = \varphi^2(u+\tau)\varphi(u+c)\varphi(u-c),$$

from which we can, by the relations (IV.3.3) and (IV.3.4), deduce (using a theorem of Euclid (Appendix B.1)) that

$$\varphi^2(u)|\varphi^2(u+\tau)|\varphi^2(u).$$

But this means (here we are using the decomposition of functions into prime factors) that $\varphi(u)|\varphi(u+\tau)|\varphi(u)$.

From the equalities

$$\varphi(u+\tau) = \psi_{u_0}(u)\varphi(u), \qquad \varphi(u) = \varphi(u+\tau)\chi_{u_0}(u),$$

where $\psi_{u_0}(u)$ and $\chi_{u_0}(u) = \psi_{u_0}^{-1}(u)$ are functions which are analytic at the point u_0 (for any $u_0 \in \mathbf{C}^p$), it follows that the function $\varphi(u+\tau)/\varphi(u)$ is an entire function without zeros, i.e., that

$$\varphi(u+\tau) = \exp[2\pi i g(u)]\varphi(u), \tag{IV.3.12}$$

where $g(u)$ is also an entire function. It is easy to see that in our case it is a polynomial of degree at most 1.

In fact, let us compare the two expressions for the quotient $\varphi(u+\tau+\Omega\varepsilon^{(h)})/\varphi(u)$ which are obtained by using the relations (IV.3.1) and (IV.3.12), first in one order and then in the other. We obtain

$$g(u+\Omega\varepsilon^{(h)}) - g(u) = \tilde{\varepsilon}^{(h)}\tilde{\Lambda}\tau + m_h, \qquad h = 1, \dots, 2p, \tag{IV.3.13}$$

where m_h are integers. Taking the partial derivatives of the functions $g(u)$ with respect to u_j, we find that each of them is a function with $2p$ independent periods $\Omega\varepsilon^{(1)}, \dots, \Omega\varepsilon^{(2p)}$. Since these functions are entire, we can conclude (by Liouville's Theorem) that $\partial g/\partial u_j$ are constant functions, i.e., that $g(u)$ must have the form

$$g(u) = \tilde{a}u + a_0, \tag{IV.3.14}$$

where a is a p-vector and a_0 is a complex number.

Now, the relations (IV.3.13) can be rewritten as follows:

$$\tilde{a}\Omega\varepsilon^{(h)} = \tilde{\varepsilon}^{(h)}\widetilde{\Omega}a = \tilde{\varepsilon}^{(h)}\widetilde{\Lambda}\tau + m_h, \qquad h = 1, \ldots, 2p;$$

or if we notice that $m_h = \tilde{\varepsilon}^{(h)}m$, as follows:

$$\widetilde{\Omega}a - \widetilde{\Lambda}\tau = m. \tag{IV.3.15}$$

We replace $\widetilde{\Omega}$ and $\widetilde{\Lambda}$ by their expressions

$$\begin{pmatrix} \pi i B^{-1} \\ A \end{pmatrix} \quad \text{and} \quad \begin{pmatrix} 0 \\ -\frac{n}{\pi i}E_p \end{pmatrix}.$$

If a $2p$-vector m is written in the form $m = \binom{\mu}{\nu}$, where μ and ν are p-vectors (with integral coordinates), then the relation (IV.3.15) can be separated into two parts:

$$\pi i B^{-1}a = \mu, \qquad Aa + \frac{n}{\pi i}\tau = \nu. \tag{IV.3.16}$$

If we solve the first for the vector a and substitute the result into the second, we find the relation

$$\begin{aligned} n\tau = \pi i\nu - AB\mu = \nu_1\beta_i\pi i B^{-1}e^{(1)} + \cdots \\ + \nu_p\beta_p B^{-1}e^{(p)} - \mu_1\beta_1 Ae^{(1)} - \cdots - \mu_p\beta_p Ae^{(p)}. \end{aligned} \tag{IV.3.17}$$

Thus we have established that $n\tau$ *is one of the periods generated by the matrix* $\Omega = (\pi i B^{-1}, A)$.

Even from this it follows that *infinitesimal periods do not exist* and therefore the function $f(u)$ constructed earlier *is nondegenerate.*

We will now show that, under some additional restrictions (not very restrictive, by the way) on $\varphi(u)$, it can be asserted that *all the integers* $\nu_i\beta_i$ and $\mu_j\beta_j$ occurring on the right-hand side of (IV.3.17) *are divisible* by n. It follows from this that the period τ itself is also generated by the matrix Ω, and this completes the proof.

According to (IV.3.12) and (IV.3.14) we have

$$\varphi(u+\tau) = \exp[2\pi i(\tilde{a}u + a_0)]\varphi(u),$$

or, if we express a in terms of μ (by the first formula (IV.3.16)) we have

$$\varphi(u+\tau) = \exp(2\tilde{\mu}B_\mu + 2\pi i a_0)\varphi(u).$$

If we replace $\varphi(u)$ by its expression (IV.3.2), in terms of theta functions, we obtain

$$\sum_{g\in\Pi} C_g \exp(-2\tilde{\mu}Bu)\Theta_n[g](u+\tau) = \exp(2\pi i a_0)\sum_{g\in\Pi} C_g\Theta_n[g](u). \tag{IV.3.18}$$

We see that $\exp(-2\tilde{\mu}Bu)\Theta_n[g](u+\tau)$ differs from $\Theta_n[g-\mu](u)$ only by a constant factor. Indeed, by (IV.2.24′)

$$\begin{aligned}
&\exp(-2\tilde{\mu}Bu)\Theta_n[g](u+\tau) \\
&\quad = \exp(-2\tilde{\mu}Bu)\sum_m \exp[2(\tilde{g}+n\tilde{m}B^{-1})B(u+\tau)+\Phi(g+nB^{-1}m)] \\
&\quad = \sum_m \exp\left[2(\tilde{g}-\tilde{\mu}+n\tilde{m}B^{-1})Bu + 2\left(\frac{\tilde{g}}{n}+\tilde{m}B^{-1}\right)\right. \\
&\qquad\qquad\qquad \left.\times\, Bn\tau + \Phi(g+nB^{-1}m)\right]. \qquad \text{(IV.3.19)}
\end{aligned}$$

Let us denote, for brevity, the last exponent by Ψ, and add to it and subtract from it $\Phi(g-\mu+nB^{-1}m)$. We find

$$\begin{aligned}
\Psi &= \{2[(\tilde{g}-\tilde{\mu})+n\tilde{m}B^{-1}]u+\Phi(g-\mu+nB^{-1}m)\} \\
&\quad - \left\{[\Phi[(g-\mu)+nB^{-1}m]-\Phi(g+nB^{-1}m)-2\left(\frac{\tilde{g}}{n}+\tilde{m}B^{-1}\right)Bn\tau\right\} \\
&= \Psi_1 - \Psi_2.
\end{aligned}$$

The first curly bracket (Ψ_1) differs from the exponent in the formula for $\Theta_n[g](u)$ (IV.2.24′) only by having g replaced by $g-\mu$. We transform the second curly bracket (Ψ_2) by using the obvious identity (see (IV.2.14))

$$\Phi(v+w)-\Phi(v) = 2\frac{\tilde{v}}{n}BABw + \Phi(w)$$

and by changing $n\tau$ into $\pi i\nu - AB\mu$ according to (IV.3.17). We obtain

$$\begin{aligned}
\Psi_2 &= -2\left(\frac{\tilde{g}}{n}+\tilde{m}B^{-1}\right)BAB\mu + \Phi(\mu) - 2\left(\frac{\tilde{g}}{n}+\tilde{m}B^{-1}\right)B(\pi i\nu - AB\mu) \\
&= \Phi(\mu) - \frac{2\pi i}{n}\tilde{g}B\nu - 2\pi i\tilde{m}\nu.
\end{aligned}$$

The last summand is an integral multiple of $2\pi i$ and can be dropped from the exponent.

As a result, (IV.3.19) takes the form

$$\begin{aligned}
\exp(-2\tilde{\mu}Bu)\Theta_n[g](u+\tau) &= \exp\left[\frac{2\pi i}{n}\tilde{g}B\nu - \Phi(\mu)\right]\sum_m \exp\{\Psi_1\} \\
&= \exp\left[\frac{2\pi i}{n}\tilde{g}B\nu - \Phi(u)\right]\Theta_n[g-\mu](u).
\end{aligned}$$

Substituting this expression into (IV.3.18), we obtain the identity

$$\sum_{g\in\Pi} C_g \exp\left(\frac{2\pi i}{n}\tilde{g}B\nu\right)\Theta_n[g-\mu](u) = \exp[2\pi i a_0 + \Phi(\mu)]\sum_{g\in\Pi} C_g\Theta_n[g](u). \qquad \text{(IV.3.20)}$$

Recall that μ is *a fixed* vector here (see the first formula (IV.3.16)) and g is *a variable* vector. Since for any $g \in \Pi$ one can find a vector $\varkappa = \varkappa(g)$ with integral coordinates such that

$$g - \mu = g' + nB^{-1}\varkappa, \quad \text{where } g' = g'(g) \in \Pi, \tag{IV.3.2.1}$$

we get

$$\Theta_n[g-\mu](u) = \Theta_n[g'](u).$$

The mapping $g \to g'$ of the parallelotope Π into itself is bijective since if $g - \mu = g' + nB^{-1}\varkappa$ and $h - \mu = g' + nB^{-1}\lambda$, where $g \in \Pi$ and $h \in \Pi$, then $h - g = nB^{-1}(\lambda - \varkappa)$ which is possible only when $\lambda = \varkappa$ and $h = g$.

We agree to take $C_{g-\mu}$ as $C_{g'}$, thus changing the index of the coefficient C_g, when g does not belong to Π, by the vector $g' \in \Pi$ which is congruent modulo nB^{-1} to this index. Then the coefficient of $\Theta_n[g-\mu](u) = \Theta_n[g'](u)$ on the left-hand side of (IV.3.20) is $C_g \exp(2\pi i\tilde{g}B\nu/n)$ and on the right-hand side it is $C_{g'}\exp[2\pi i a_0 + \Phi(\mu)] = C_{g'}\rho$, where ρ does not depend on g. Changing $C_{g'}$ into $C_{g-\mu}$ (according to the agreement made above) we obtain

$$C_g \exp\left(\frac{2\pi i}{n}\tilde{g}B\nu\right) = C_{g-\mu}\rho. \tag{IV.3.22}$$

There are two possibilities for the relation between g and $g'(g)$. Either $g'(g) = g$ (if equality holds for one value of g, then $\mu = nB^{-1}(-\varkappa)$ and therefore it holds for all $g \in \Pi$) or $g'(g) \neq g$ for all $g \in \Pi$. We believe that the first case suits our purpose better.

In fact, if $g'(g) = g$, i.e., if $C_g = C_{g-\mu}$, then as we have already noticed, $\mu = nB^{-1}(-\varkappa)$. This means that the following relation holds for the coordinates μ_j of the vector μ:

$$\mu_j = -\frac{n}{\beta_j}\varkappa_j, \qquad \varkappa_j \text{ are integers}, \qquad j = 1, \dots, p,$$

and therefore all the coefficients $\mu_j\beta_j$ in (IV.3.17) are divisible by n.

We require the coefficients C_g to satisfy the condition $C_g \neq 0$ for all $g \in \Pi$. *By doing this we somewhat restrict the choice of the Jacobian function* $\varphi(u)$ (IV.3.2).

Then it follows from the formulas (IV.3.22) that

$$\exp\left(\frac{2\pi i}{n}\tilde{g}B\nu\right) = \rho$$

for all $g \in \Pi$. Setting $g = 0$, we obtain $\rho = 1$. Now let $g = e^{(j)}$ ($j = 1, \dots, p$). Then it must be true that

$$\exp\left(\frac{2\pi i}{n}\tilde{e}^{(j)}B\nu\right) = 1,$$

i.e., all the numbers $\nu_j\beta_j$, $j = 1, \ldots, p$, are also multiples of n.[22] Thus *if all the coefficients* C_g *in* (IV.3.2) *are different from zero and the mapping* $g \to g'$ *defined by* (IV.3.21) *is the identity, then the period* τ *(of the functions* (IV.3.7)) *is generated by the matrix* Ω.

It remains to prove that this case is *typical* in the sense that *the mapping* (IV.3.21) *is the identity for any period* τ *of the function* $f(u)$, *provided that the coefficients* C_g (we still consider them to be nonzero) *are not restricted by relations of a very special type.*

Indeed, assume that the mapping (IV.3.21) is not the identity for a given function $\varphi(u)$ and a given period τ corresponding to the function (IV.3.7). Then we write the relations (IV.3.22) for the indices $g, g-\mu, g-2\mu, \ldots, g-(n-1)\mu$ and multiply their corresponding sides. We obtain

$$C_g C_{g-\mu} \cdots C_{g-(n-1)\mu} \exp\left[2\pi i \left(\tilde{g} - \frac{n-1}{2}\tilde{\mu}\right) B\nu\right]$$
$$= C_{g-\mu} \cdots C_{g-(n-1)\mu} C_{g-n\mu} \rho^n .$$

Since $g-(g-n\mu) = n\mu = nB^{-1}(B\mu) = nB^{-1}\varkappa$, where $\varkappa$ is a vector with integral coordinates, $C_g = C_{g-n\mu}$ and therefore the preceding equality yields

$$\rho^n = \exp[2\pi i\tilde{g}B\nu - \pi i(n-1)\tilde{\mu}B\nu] = \pm 1 .$$

But the quotient $C_{g-\mu}/C_g$ of the coefficients for different (under our assumption) theta functions $\Theta_n[g-\mu](u)$ and $\Theta_n[g](u)$ is equal to

$$\chi(g) = \frac{1}{\rho}\exp\left(\frac{2\pi i}{n}\tilde{g}B\nu\right)$$

and therefore must satisfy the condition

$$\chi(g)^n = \pm 1, \quad \text{i.e., } \chi^{2n}(g) = 1 .$$

Hence it is enough to demand that *for no pair of coefficients* C_g *and* C_{g^*} *with different indices* g *and* g^* *may the ratio* C_{g^*}/C_g *be equal to one of the* $2n$*th roots of unity. Then the case* $g' \neq g$ *will be impossible.*

Thus in order for the function (IV.3.7) to be an Abelian function generated by a given matrix Ω, it is enough to demand that the corresponding Jacobian function $\varphi(u)$, defined by the formula (IV.3.2), satisfies the following two conditions:

(1) $C_g \neq 0$ for all $g \in \Pi$;

(2) $[C_{g^*}/C_g]^{2n} \neq 1$ for all pairs g^*, g such that $g^* \neq g$, $g^* \in \Pi$, $g \in \Pi$.

Since these conditions can be satisfied (in general) in infinitely many ways, we *obtain an infinite set of Abelian functions for each of which a given matrix* Ω *generates its set of periods.*

---

(22) If $l = 1$ and $n = \beta_p = \beta_{p-1} = \cdots = \beta_q$, then $e^{(p)}, \ldots, e^{(q)}$ does not belong to Π; however, the numbers $\nu_p\beta_p, \ldots, \nu_q\beta_q$ are still multiples of n.

It follows from our discussion that the conclusion also holds when an Abelian function has other periods besides those generated by the matrix Ω. In particular, the degenerate functions can also be represented in the form of quotients of two Jacobian functions of the same type.

An Abelian function whose periods coincide exactly with the set generated by the matrix Ω will be said to *correspond properly to the Riemann matrix* Ω. It is obvious that if $f_1(u)$ and $f_2(u)$ are two Abelian functions corresponding properly to Ω, then for any constants c_1 and c_2 the functions $c_1 f_1(u) + c_2 f_2(u)$, as well as $f_1(u) f_2(u)$ and $f_1(u)/f_2(u)$, are meromorphic functions, the set of whose periods contains the set of periods generated by Ω. However, these functions can also have other periods. In particular, the degenerate functions and constant functions (which form a subset of the degenerate functions) can be obtained in this way. Hence it is useful to consider, along with the set of all Abelian functions corresponding properly to a Riemann matrix Ω, the larger set of periodic meromorphic functions whose set of periods includes all periods generated by Ω.

Thus *the set of all meromorphic functions of variables* $u_1, \dots, u_p$ *such that each of these functions possesses all of the periods generated by a given Riemann matrix* Ω *is a field. This field is called the field of Abelian functions corresponding to* Ω.

We note that the field of Abelian functions which correspond to a Riemann matrix Ω always includes, together with each function $f(u)$, its partial derivatives $\partial f/\partial u_1, \dots, \partial f/\partial u_p$. In fact, all these derivatives are meromorphic functions and each period of a function $f(u)$ is also a period of any of its derivatives.

We introduced in section III.1 the notion of an equivalence class for the set of Riemann matrices: any matrix Ω^* in the class of Ω is obtained from Ω by a transformation of the form $\Omega^* = A\Omega M$, where M is an integral unimodular $(2p, 2p)$ matrix and A is any nonsingular (p, p) matrix with integral coefficients. The matrix M guarantees that the transformation from one system of principal periods to another does not change the set of periods. The matrix A defines a linear transformation of the variables $u^* = Au$. Naturally the fields of Abelian functions corresponding to the matrices Ω and ΩM are not different. But we can go even further and can say that the fields which are obtained from each other by means of a linear change of variables $u^* = Au$ are fields equivalent to each other. Then *there is a one-to-one correspondence between the set of all equivalence classes of Riemann* $(p, 2p)$ *matrices and the set of equivalence classes of the fields of Abelian functions of* p *variables.*

§4. Theta functions with characteristics

Since $n = l\beta_p$, the values $\beta_p = \cdots = \beta_1 = 1$ correspond to the case $n = 1$, i.e., the matrix B becomes E_p. Hence the matrices Ω, Λ and the parametric vector γ have the following forms:

$$\Omega = (\pi i E_p, A), \qquad \Lambda = \left(0, -\frac{1}{\pi i}E_p\right), \qquad \gamma = \begin{pmatrix} 0 \\ -\frac{1}{2\pi i} \ \operatorname{sp} A \end{pmatrix}, \tag{IV.4.1}$$

where $\operatorname{sp} A$ denotes the p-vector formed by all diagonal elements of the matrix A.

In this case the parallelotope Π contains only one point, namely the zero vector. Hence there is a single theta function of the first order which, according to the formula (IV.2.24′), has the form

$$\Theta_1[0](u) = \sum_m \exp[2\tilde{m}u + \Phi(m)],$$

where $\Phi(m) = \tilde{m}Am$. This is called *the general theta function of the first order* and is denoted by $\vartheta(u) = \vartheta(u; A)$. Thus

$$\vartheta(u; A) = \sum_m \exp[\tilde{m}(Am + 2u)]. \tag{IV.4.2}$$

Since the matrix $A = (\alpha_{jk})$ is symmetric, the theta function $\vartheta(u; A)$ depends on $\frac{1}{2}p(p+1)$ complex parameters α_{jk} $(j \leq k)$. The numbers α_{jk} are called *the moduli of the theta function* $\vartheta(u; A)$. Recall that they satisfy certain inequalities which follow from the relation $\tilde{u}A'\bar{u} < 0$ when $u \neq 0$ $(A' = \operatorname{Re} A)$. The variables $u_1, \ldots, u_p$ are called *the arguments of the theta function.* Because $\vartheta(u; A)$ is *unique* for a given matrix A, *any Jacobian function* $\varphi(u)$ *of type* $\{\Omega, \Lambda, \gamma\}$ can be derived from it by multiplication by a complex constant. Starting from $\vartheta(u; A)$ we can obtain an infinite set of other Jacobian functions of type $\{\Omega, \Lambda, \gamma^*\}$, where γ^* is a parametric vector distinct from γ.

At the end of section IV.1 it was shown that if $\varphi(u)$ is a Jacobian function of type $\{\Omega, \Lambda, \gamma\}$, the function

$$\varphi^*(u) = \exp[\pi i(\tilde{u}\Sigma u + 2\tilde{\alpha}u)]\varphi(u + c),$$

where Σ is any symmetric (p, p) matrix, is a Jacobian matrix of the type $\{\Omega, \Lambda + \Sigma\Omega, \gamma^*\}$. Here the new parametric vector γ^* is computed by the formula (IV.1.22) as follows:

$$\gamma^* = \gamma + \frac{1}{2}\zeta + \tilde{\Omega}\alpha + \tilde{\Lambda}c,$$

where $\zeta = \operatorname{sp}(\tilde{\Omega}\Sigma\Omega)$.

Because we want to leave the second period matrix Λ unchanged we set $\Sigma = 0$ and then we have $\zeta = 0$. Thus we define the Jacobian function

$$\varphi^*(u) = \exp(2\pi i\tilde{\alpha}u)\vartheta(u + c; A). \tag{IV.4.3}$$

In order to arrive at the notation of the classical theory of theta functions we also set

$$\alpha = \frac{1}{\pi i}g, \qquad c = Ag + \pi i h,$$

where g and h are two p-vectors with any complex components. By what we said above, the function $\varphi^*(u)$ must be of type $\{\Omega, \Lambda, \gamma^*\}$, where the matrices Ω and Λ have the form (IV.4.1) and the $2p$-vector γ^* is given by the formula

$$\begin{aligned}\gamma^* &= \gamma + \frac{1}{\pi i}\widetilde{\Omega} g + \widetilde{\Lambda}(Ag + \pi i h)\\ &= \begin{pmatrix} 0 \\ -\frac{1}{2\pi i}\operatorname{sp} A\end{pmatrix} + \begin{pmatrix} g \\ \frac{1}{\pi i} Ag\end{pmatrix} + \begin{pmatrix} 0 \\ -\frac{1}{\pi i}Ag - h\end{pmatrix}\\ &= \begin{pmatrix} g \\ -\frac{1}{2\pi i}\operatorname{sp} A - h\end{pmatrix}. \qquad \text{(IV.4.4)}\end{aligned}$$

By definition, *the general theta function with characteristics* g *and* h is a Jacobian function differing from (IV.4.3) only by a numerical normalizing factor $\exp[\tilde{g}(Ag + 2\pi i h)]$, i.e.,

$$\vartheta\begin{bmatrix} g \\ h\end{bmatrix}(u;\, A) = \exp[\tilde{g}(Ag + 2u + 2\pi i h)]\vartheta(u + Ag + \pi i h;\, A). \qquad \text{(IV.4.3}'\text{)}$$

From the formula (IV.4.2), we obtain the Fourier series

$$\vartheta\begin{bmatrix} g \\ h\end{bmatrix}(u;\, A) = \sum_m \exp\{(\tilde{m} + \tilde{g})[A(m + g) + 2u + 2\pi i h]\}. \qquad \text{(IV.4.5)}$$

It is obvious that when $g = h = 0$ we again find the general theta function $\vartheta(u;\, A)$ of the first order:

$$\vartheta\begin{bmatrix} 0 \\ 0\end{bmatrix}(u;\, A) = \vartheta(u;\, A). \qquad \text{(IV.4.6)}$$

The fact that $\vartheta[{}^g_h](u;\, A)$ is of type $\{\Omega, \Lambda, \gamma^*\}$, where Ω and Λ are defined by (IV.4.1) and γ^* is defined by (IV.4.4), can be expressed by means of the identities:

$$\vartheta\begin{bmatrix} g \\ h\end{bmatrix}(u + \pi i e^{(k)};\, A) = \exp(2\pi i \tilde{e}^{(k)} g)\vartheta\begin{bmatrix} g \\ h\end{bmatrix}(u;\, A), \qquad \text{(IV.4.7}'\text{)}$$

$$\vartheta\begin{bmatrix} g \\ h\end{bmatrix}(u + Ae^{(k)};\, A) = \exp\{-\tilde{e}^{(k)}[2u + \operatorname{sp} A + 2\pi i h]\}\vartheta\begin{bmatrix} g \\ h\end{bmatrix}(u;\, A). \qquad \text{(IV.4.7}''\text{)}$$

Applying them again, we obtain the identity

$$\vartheta\begin{bmatrix} g \\ h\end{bmatrix}(u + A\varkappa + \pi i\lambda;\, A) = \exp[-\tilde{\varkappa}(2u + A\varkappa) + 2\pi i(\tilde{\lambda} g - \tilde{\varkappa} h)]\vartheta\begin{bmatrix} g \\ h\end{bmatrix}(u;\, A), \qquad \text{(IV.4.8)}$$

where $\varkappa$ and λ are any p-vectors with integral coordinates. This formula can also be obtained directly from the expansion (IV.4.5).

Using the expansion (IV.4.5) we easily check the truth of the general formula

$$\begin{aligned}\begin{bmatrix} g + g' \\ h + h'\end{bmatrix}(u;\, A) = {}& \exp[\tilde{g}'(Ag' + 2u + 2\pi i h + 2\pi i h')]\\ & \times \vartheta\begin{bmatrix} g \\ h\end{bmatrix}(u + Ag' + \pi i h';\, A). \qquad \text{(IV.4.9)}\end{aligned}$$

Using (IV.4.8), when $g' = \varkappa$ and $h' = \lambda$ are vectors with integral coordinates, we obtain

$$\vartheta\begin{bmatrix} g+\varkappa \\ h+\lambda \end{bmatrix}(u;A) = \exp(2\pi i\tilde{\lambda}g)\vartheta\begin{bmatrix} g \\ h \end{bmatrix}(u;A). \tag{IV.4.9$'$}$$

We also notice the following property of the theta functions with characteristics:

$$\vartheta\begin{bmatrix} g \\ h \end{bmatrix}(-u;A) = \vartheta\begin{bmatrix} -g \\ -h \end{bmatrix}(u;A). \tag{IV.4.10}$$

Indeed, the decomposition (IV.4.5) gives

$$\begin{aligned}\vartheta\begin{bmatrix} g \\ h \end{bmatrix}(-u,A) &= \sum_m \exp\{(\tilde{m}+\tilde{g})[A(m+g) - 2u + 2\pi ih]\} \\ &= \sum_m \exp\{(-\tilde{m}-\tilde{g})[A(-m-g) + 2u - 2\pi ih]\}\end{aligned}$$

from which it then follows that (IV.4.10) is true provided one changes m to $-m$. In particular, when $g = h = 0$ we obtain, by (IV.4.10),

$$\vartheta(-u;A) = \vartheta(u,A), \tag{IV.4.10$'$}$$

i.e., *the general theta function is even.*

Now consider any number n of theta functions of the first order with the same moduli (i.e., with the same matrix A) but with different (in general) characteristics

$$\vartheta\begin{bmatrix} g_1 \\ h_1 \end{bmatrix}(u;A), \ldots, \vartheta\begin{bmatrix} g_n \\ h_n \end{bmatrix}(u;A). \tag{IV.4.11}$$

Since they are all Jacobian functions corresponding to the same matrix $\Omega = (\pi iE, A)$, their product $\varphi(u)$ is also a Jacobian function corresponding to the same matrix (section III.4). More precisely, by multiplying the corresponding sides of the formulas (IV.4.7$'$) and (IV.4.7$''$) we obtain

$$\begin{gathered}\varphi(u+\pi ie^{(k)}) = \exp(2\pi i\tilde{e}^{(k)}g)\varphi(u), \\ \varphi(u+Ae^{(k)}) = \exp[-\tilde{e}^{(k)}(2nu + n\,\mathrm{sp}\,A + 2\pi ih)]\varphi(u),\end{gathered}$$

where $g = \Sigma_{j=1}^n g_j$ and $h = \Sigma_{j=1}^n h_j$. These formulas express the fact that the Jacobian function $\varphi(u)$ is of the type

$$\{\Omega, n\Lambda, \gamma^*\}, \quad \text{where } \gamma^* = \begin{pmatrix} g \\ -\frac{n}{2\pi i}\,\mathrm{sp}\,A - h \end{pmatrix}.$$

A function with these properties is called *a theta function of order n with characteristics g and h.*

Consequently the number of such functions which are linearly independent of each other can be equal to $|\delta| = n^p$ for given g and h. In the case when $g = h = 0$ these functions are the theta functions of order n which we already know:

$$\Theta_n[\varkappa](u;A) = \sum_m \exp\left\{\left(\tilde{m} + \frac{1}{n}\tilde{\varkappa}\right) n \left[A\left(m + \frac{1}{n}\varkappa\right) + 2u\right]\right\}, \tag{IV.4.12}$$

where the vector $\varkappa$ can span all the points of the parallelotope Π, determined in this case ($\beta_1 = \cdots = \beta_p = 1$) by the inequalities $0 \le \varkappa_j < n$ ($j = 1, \dots, p$). Comparing this formula with (IV.4.5), we conclude that

$$\Theta_n[\varkappa](u; A) = \vartheta \begin{bmatrix} \frac{\varkappa}{n} \\ 0 \end{bmatrix} (nu; nA). \tag{IV.4.13}$$

Furthermore, (IV.4.3′) allows us to express the theta function of order n that is under consideration in terms of the general theta function of the first order:

$$\Theta_n[\varkappa](u; A) = \exp\left[\tilde{\varkappa}\left(A\frac{\varkappa}{n} + 2u\right)\right] \vartheta(nu + A\varkappa; nA). \tag{IV.4.14}$$

In a way similar to that by which the functions of the first order with characteristics g and h were obtained from the functions $\vartheta(u; A)$ by using transformations of the form (IV.4.3), the theta functions with any characteristics g and h and of the same order as $\Theta_n[\varkappa](u; A)$ can be obtained. They are defined by the formula

$$\begin{aligned}\Theta_n \begin{bmatrix} g \\ h \end{bmatrix}, \varkappa \Big] (u; A) &= e^{2\tilde{g}u}\Theta_n[\varkappa]\left(u + \frac{1}{n}Ag + \frac{\pi i}{n}h; A\right) \\ &= e^{2\tilde{g}u}\vartheta \begin{bmatrix} \frac{\varkappa}{n} \\ 0 \end{bmatrix} (nu + Ag + \pi i h; nA). \end{aligned} \tag{IV.4.15}$$

The function $\Theta\left[\begin{smallmatrix} g \\ h \end{smallmatrix}, \varkappa\right](u; A)$ satisfies the definition of the theta function of order n with characteristics when the vector $\varkappa$ has nonnegative integral coordinates, each of which is less than n. That is, this function is a Jacobian function of the type $\{\Omega, n\Lambda, \gamma^*\}$ where the parametric vector is given by

$$\gamma^* = \begin{pmatrix} -g \\ -\frac{n}{2\pi i}\operatorname{sp} A - h \end{pmatrix}.$$

We transform the right-hand side of (IV.4.15) by applying (IV.4.9), which we read from the right to left. Here we change g to $\varkappa/n$, h to 0, u to nu, A to nA, g' to g/n, and finally h' to h in the latter formula. We obtain

$$\begin{aligned}&\vartheta \begin{bmatrix} \varkappa/n \\ 0 \end{bmatrix} (nu + Ag + \pi i h; nA) \\ &= \exp\left[-\frac{\tilde{g}}{n}\left(nA\frac{g}{n} + 2nu + 2\pi i h\right)\right] \vartheta \begin{bmatrix} \frac{\varkappa + g}{n} \\ h \end{bmatrix} (nu; nA). \\ &= \exp\left[-\frac{1}{n}\tilde{g}(Ag + 2\pi i h) - 2\tilde{g}u\right] \vartheta \begin{bmatrix} \frac{\varkappa + g}{n} \\ h \end{bmatrix} (nu; nA). \end{aligned}$$

Hence (IV.4.15) can be expressed in the form

$$\Theta_n \begin{bmatrix} g \\ h \end{bmatrix}, \varkappa \Big] (u; A) = \exp\left[-\frac{1}{n}\tilde{g}(Ag + 2\pi i h)\right] \vartheta \begin{bmatrix} \frac{\varkappa + g}{n} \\ h \end{bmatrix} (nu; nA). \tag{IV.4.16}$$

The most general theta function of order n with characteristics g and h can be represented as a linear combination of the ones above with arbitrary constant coefficients:

$$\Theta_n\begin{bmatrix} g \\ h \end{bmatrix}(u; A) = \sum_{\varkappa} C_{\varkappa}\vartheta\begin{bmatrix} \frac{\varkappa+g}{n} \\ h \end{bmatrix}(nu; nA), \tag{IV.4.17}$$

where the summation is taken over all vectors $\varkappa$, with integer coordinates that satisfy the conditions

$$0 \le \varkappa_j < n, \qquad j = 1, \dots, p.$$

We illustrate the facts about theta functions discussed above by looking at the case $p = 1$, when the functions are *elliptic theta functions*. When $p = 1$ the matrices Ω and Λ are reduced to the forms $\Omega = (\pi i, \alpha)$ and $\Lambda = (0, -1/\pi i)$. If $\alpha = \alpha' + i\alpha''$ here, the number α' must be negative: $\alpha' < 0$ (this is implied by the Riemann inequality).

The general theta function of the first order is defined here by the expansion

$$\vartheta(u; \alpha) = \sum_m \exp(\alpha m^2 + 2mu),$$

where the summation is over all integers m. In order to use the notation which is generally accepted in the theory of elliptic theta functions we need to set

$$u = \pi i v, \qquad \alpha = \pi i \omega\left(\operatorname{Im}\omega = -\frac{1}{\pi}\alpha' > 0\right), \qquad q = e^{\alpha} = e^{\pi i \omega}.$$

We get

$$\vartheta(u; \alpha) = \sum_m q^{m^2} e^{2m\pi i v} = 1 + 2\sum_1^{\infty} q^{m^2} 2m\pi v.$$

This function, as we know, is a theta function with zero characteristics $\vartheta\left[\begin{smallmatrix} 0 \\ 0 \end{smallmatrix}\right](u; \alpha)$. According to Jacobi's notation it is written in the form

$$\vartheta_3(v, q) = \vartheta\begin{bmatrix} 0 \\ 0 \end{bmatrix}(u, \alpha) = 1 + 2\sum_1^{\infty} q^{m^2}\cos 2m\pi v.$$

Three other Jacobi theta functions are functions with fractional characteristics:

$$\vartheta_0(v, q) = \vartheta\begin{bmatrix} 0 \\ 1/2 \end{bmatrix}(u; \alpha)$$

(we denoted it by $\vartheta_4(v, q)$ in section I.3), and

$$\vartheta_1(v; q) = -\vartheta\begin{bmatrix} 1/2 \\ 1/2 \end{bmatrix}(u; \alpha), \qquad \vartheta_2(v, q) = \vartheta\begin{bmatrix} 1/2 \\ 0 \end{bmatrix}(u; \alpha).$$

The general formula (IV.4.3′) takes the form

$$\vartheta\begin{bmatrix} g \\ h \end{bmatrix}(u, \alpha) = \exp[\tilde{g}(\alpha g + 2u + 2\pi i h)]\vartheta(u + \alpha g + \pi i h; \alpha),$$

and therefore

$$\begin{aligned}
\vartheta_0(v\,;q) &= \vartheta\left(u+\frac{1}{2}\pi i\,;\alpha\right) \\
&= 1 - 2q\cos 2\pi v + 2q^4\cos 4\pi v - 2q^9\cos 6\pi v + \cdots, \\
\vartheta_1(v\,;q) &= -i\exp\left(u+\frac{1}{4}\alpha\right)\vartheta\left(u+\frac{1}{2}\alpha+\frac{1}{2}\pi i\,;\alpha\right) \\
&= 2q^{1/4}\sin \pi v - 2q^{9/4}\sin 3\pi v + 2q^{28/4}\sin 5\pi v - \cdots, \\
\vartheta_2(v\,;q) &= \exp\left(u+\frac{1}{4}\alpha\right)\vartheta\left(u+\frac{1}{2}\alpha\,;\alpha\right) \\
&= 2q^{1/4}\cos \pi v + 2q^{9/4}\cos 3\pi v + 2q^{28/4}\cos 5\pi v + \cdots.
\end{aligned}$$

The square of a theta functions is a theta function of the second order, and moreover it has integral characteristics and therefore can be reduced to a function with zero characteristics. But the latter can be expressed, in the case of order $n = 2$, in the form of linear combinations of $n^p = 2$ theta functions of the first order ($\vartheta\left[\begin{smallmatrix}\varkappa/2\\0\end{smallmatrix}\right](2u;\,2\alpha)$). Therefore *the squares of any three elliptic theta functions must be related by a linear relation with constant coefficients.* For example,

$$C_0\vartheta_0^2(v\,;q) + C_1\vartheta_1^2(v\,;q) + C_2\vartheta_2^2(v\,,q) = 0\,.$$

Setting $v = 0$ here and denoting $\vartheta_0(0;\,q)$ by ϑ_0 and $\vartheta_2(0;\,q)$ by ϑ_2 (here $\vartheta_1(0;\,q) = 0$), we get $C_0\vartheta_0^2 + C_2\vartheta_2^2 = 0$.

Similarly, when $v = 1/2$ we find that $\vartheta_0(1/2;\,q) = \vartheta_3$, $\vartheta_1(1/2;\,q) = \vartheta_2$, $\vartheta_3(1/2;\,q) = 0$, and therefore

$$C_0\vartheta_3^2 + C_1\vartheta_2^2 = 0\,.$$

Thus $C_0/\vartheta_2^2 = C_1/(-\vartheta_3^2) = C_2/(-\vartheta_0^2)$, from which it follows that the relation we are looking for among the squares of theta functions has the form

$$\vartheta_2^2\vartheta_0^2(v\,;q) = \vartheta_3^2\vartheta_1^2(v\,;q) + \vartheta_0^2\vartheta_2^2(v\,;q)\,.$$

§5. Weierstrass-Poincaré theorem

The importance of elliptic functions in Algebraic Geometry (geometry of algebraic curves) is due to the fact that any algebraic curve of genus one can be *parameterized* by elliptic functions. That is, if $f(z\,,w) = 0$ is an irreducible algebraic equation of genus 1 (its degree with respect to w can be arbitrarily high), one can always find elliptic functions $z = \varphi(t)$ and $w = \psi(t)$ with a common fundamental parallelogram of periods Π_0 such that these functions satisfy the given algebraic equation for all t. In other words, *a multiple-valued algebraic relation* $w = w(z)$ (where there is in general the same number $n \geq 3$ of different values of w for each z) can be represented parametrically in terms of two *single-valued* meromorphic (elliptic) functions $z = \varphi(t)$ and $w = \psi(t)$.

A classical example of this is provided by an algebraic curve of order three: $w^2 = 4z^3 - g_2 z - g_3$, where the complex numbers g_2 and g_3 must satisfy only the single condition $27g_3^2 + g_2^3 \neq 0$. In the Weierstrass theory of elliptic functions it is proved that under the preceding condition we can find the periods $2\omega_1$ and $2\omega_3$ for which $\operatorname{Im} \omega_3/\omega_1 > 0$ (they are the roots of a system of transcendental equations $g_2 = 60 \sum'_{m_1, m_2} 1/(2m_1\omega_1 + 2m_2\omega_3)^4$, $g_3 = 140 \sum'_{m_1, m_2} 1/(2m_1\omega_1 + 2m_2\omega_3)^6$), such that the elliptic functions $\wp(t)$ and $\wp'(t)$ for which these periods are basic provide the uniformization of the curve: it suffices to set $z = \wp(t)$ and $w = \wp'(t)$.

Poincaré showed that the uniformization of any algebraic function of genus $p > 1$, where the function is defined by an irreducible algebraic equation $f(z, w) = 0$, can be achieved with two automorphic functions with respect to some properly discontinuous group of linear fractional transformations

$$\Gamma \left\{ T_n(t) = \frac{a_n t + b_n}{c_n t + d_n}, \ a_n d_n - b_n c_n = 1, \ T_0(t) = t \right\}.$$

These functions make the given algebraic equation an identity. They are single-valued and meromorphic on the unit disk $|t| < 1$. Moreover, they do not change if one substitutes

$$T_n(t) \cdot \varphi[T_n(t)] = \varphi(t), \qquad \psi[T_n(t)] = \psi(t)$$

for t (*this is a generalization of the property of double periodicity*).([23])

The role of Abelian functions in algebraic geometry of multidimensional manifolds can be compared to the role of elliptic (but not of general automorphic) functions in the theory of algebraic curves. Using Abelian functions one can also achieve uniformization of and therefore study a wide class of algebraic varieties (but not the most general varieties). These special varieties are called *Abelian varieties.*

We refer the reader to the specialized literature and especially to the second part of the book by F. Conforto (see the footnote on page 44) to whose exposition ours is very similar. Here we will consider the question about algebraic relations between the theta functions and Abelian functions corresponding to some matrix Ω.

We begin with preparations for the proof of a theorem that states that there is always an algebraic relation among any $p + 1$ Abelian functions belonging to the same field $\mathfrak{K}$. This theorem was actually stated by Weierstrass, who used it in his arguments but who never published a proof. The first published proof was given by Poincaré in 1897.

First we will prove that *there is a homogeneous algebraic relation among any* $p + 2$ *nondegenerate Jacobian functions which belong to the same type.*

Let Ω be a Riemann matrix of order p, let $\mathfrak{K}$ be the field of Abelian functions of p variables which correspond to Ω, and let $\varphi(u)$ be some

([23])L. R. Ford, *Automorphic functions*, McGraw-Hill, New York, 1929.

nondegenerate Jacobian function of type $\{\Omega, \Lambda, \gamma\}$. We know that there are $|\delta| = \sqrt{|N|}$ linearly independent functions of the same type as $\varphi(u)$ where $N = \widetilde{\Omega}\Lambda - \widetilde{\Lambda}\Omega$. All other theta functions of the same type are linear combinations of these (section IV.2). Notice that for two Jacobian functions $\varphi_1(u)$ and $\varphi_2(u)$ of types $\{\Omega, \Lambda_1, \gamma^{(1)}\}$ and $\{\Omega, \Lambda_2, \gamma^{(2)}\}$ we have

$$\varphi_1(u+\Omega\varepsilon^{(h)})\varphi_2(u+\Omega\varepsilon^{(h)}) = \exp\{2\pi i\varepsilon^{(h)}[(\widetilde{\Lambda}_1+\widetilde{\Lambda}_2)u+\gamma^{(1)}+\gamma^{(2)}]\}\varphi_1(u)\varphi_2(u).$$

In other words, *the product* $\varphi_1(u)\varphi_2(u)$ *of two Jacobian functions of types* $\{\Omega, \Lambda_1, \gamma^{(1)}\}$ *and* $\{\Omega, \Lambda_2, \gamma^{(2)}\}$ *is a Jacobian function corresponding to the same Riemann matrix* Ω *but with the second period matrix* $\Lambda_1+\Lambda_2$ *and the parametric vector* $\gamma^{(1)}+\gamma^{(2)}$.

Applying the same reasoning to the product of l Jacobian functions *all of the same type* $\{\Omega, \Lambda, \gamma\}$ (where some of the functions may be equal) we obtain a Jacobian function of the type $\{\Omega, l\Lambda, l\gamma\}$. Its characteristic matrix is $\widetilde{\Omega}l\Lambda - l\widetilde{\Lambda}\Omega = lN$ and the "determinant" is equal to $\sqrt{|lN|} = l^p\sqrt{|N|} = l^p\delta$.

Therefore *there are* $l^p|\delta|$ *linearly independent Jacobian functions of this type and the number* $l^p|\delta|$ *is the largest possible.*

Now consider $p+2$ Jacobian functions of the same type $\{\Omega, \Lambda, \gamma\}$ with "determinant" $|\delta| \geq 1$. We will not assume that they are linearly independent or even that they are distinct. Denote these functions by

$$x_0 = \varphi(u), x_1 = \varphi(u), \dots, x_{p+1} = \varphi_{p+1}(u).$$

The number of products of l such functions (possibly with repeated factors) is $(l+p+1)!/(p+1)!l!$ and all these products are Jacobian functions of the same type $\{\Omega, l\Lambda, l\gamma\}$. Among these, there are not more than $l^p|\delta|$ which are linearly dependent. But this last number is obviously (for sufficiently large l) less than

$$\frac{(l+p+1)!}{(p+1)!l!} = \frac{(l+p+1)(l+p)\cdots(l+1)}{p!},$$

because the last expression is a polynomial in l (when p is fixed) of degree $p+1$, i.e., of degree greater than p. Hence the set of monomials of the form $\varphi_0^{\alpha_0}\cdots\varphi_{p+1}^{\alpha_{p+1}}$, where $\alpha_0+\cdots+\alpha_{p+1} = l$, must contain linearly dependent functions. In other words, there is a homogeneous polynomial $\Phi(x_0, x_1, \dots, x_{p+1})$ of degree l whose value is 0 for a sufficiently large l provided that one substitutes for each x_j a corresponding Jacobian function $\varphi_j(u)$ $(j = 0, 1, \dots, p+1)$, i.e.,

$$\Phi[\varphi_0(u), \dots, \varphi_{p+1}(u)] = 0 \qquad \text{(IV.5.1)}$$

(an identity with respect to all $u_1, \dots, u_p$). The coefficients of this polynomial are complex numbers. It is obvious that Φ can be considered to be an irreducible polynomial over the field $\mathbf{C}^p$ by substituting for Φ, if necessary, the corresponding irreducible factor of the product determining Φ.

Thus we have proved the following important theorem.

THEOREM. *At least one homogeneous algebraic relation holds among any* $p+2$ *nondegenerate Jacobian functions of* p *complex variables* $u_1, \ldots, u_p$ *of the same type* $\{\Omega, \Lambda, \gamma\}$.

We notice that this theorem is still true when there are degenerate functions among $\varphi_0, \ldots, \varphi_{p+1}$. Indeed, in that case one can find a nondegenerate Jacobian function ψ of another type such that all products $\varphi_0\psi, \ldots, \varphi_{p+1}\psi$ are also nondegenerate functions of the same type. Then it remains only to divide (using the fact that the relation is homogeneous) all terms by a common factor of the form ψ^l.

A corresponding assertion for Abelian functions follows from the theorem proved above. Let some $p+1$ Abelian functions $f_1(u), f_2(u), \ldots, f_{p+1}(u)$ (some of which may be degenerate) correspond to the same Riemann matrix Ω. In view of the main theorem about the representation of Abelian functions in the form of a quotient of two Jacobian functions, it can be asserted that there are nondegenerate Jacobian functions $\varphi_0(u), \ldots, \varphi_{p+1}(u)$ belonging to the same type such that

$$f_1(u) = \frac{\varphi_1(u)}{\varphi_0(u)}, \ldots, f_{p+1}(u) = \frac{\varphi_{p+1}(u)}{\varphi_0(u)}$$

(all the fractions above can be reduced to a common denominator by multiplying by the corresponding nondegenerate Jacobian functions).

Because of the theorem proved above, there exists a homogeneous polynomial $\Phi(x_0, \ldots, x_p, x_{p+1})$ of sufficiently high degree l such that

$$\Phi(x_0, x_1, \ldots, x_{p+1}) = 0$$

becomes an identity in $u_1, \ldots, u_p$ if one makes the substitution $x_j = \varphi_j(u)$ $(j = 0, 1, \ldots, p+1)$. Dividing all terms of Φ by x_0^l, we obtain a polynomial $F(y_1, \ldots, y_{p+1})$ which is nonhomogeneous with respect to the variables $y_1 = x_1/x_0, \ldots, y_{p+1} = x_{p+1}/x_0$ such that the equation $F(y_1, \ldots, y_{p+1}) = 0$ becomes an identity with respect to the variables $u_1, \ldots, u_p$ if one sets $y_j = \varphi_j(u)/\varphi_0(u) = f_j(u)$ $(j = 1, \ldots, p+1)$.

Thus we have proved the following:

WEIERSTRASS-POINCARÉ THEOREM. *Among any* $p+1$ *Abelian functions* $f_1(u), \ldots, f_{p+1}(u)$ *belonging to the same field of Abelian functions, there always exists an algebraic relation of the form*

$$F[f_1(u), \ldots, f_{p+1}(u)] \equiv 0.$$

A nonhomogeneous polynomial $F(y_1, \ldots, y_{p+1})$ *can be considered to be irreducible.*

As a particular case of the Weierstrass-Poincaré theorem (when $p = 1$) we get the following classical theorem:

THEOREM OF WEIERSTRASS. *Between any two elliptic functions $f_1(u)$ and $f_2(u)$ which have periods of the form $2m_1\omega_1 + 2m_2\omega_3$ (m_1 and m_2 run through all possible integers and moreover $\operatorname{Im}(\omega_3/\omega_1) > 0$; also the functions $f_1(u)$ and $f_2(u)$ might have some other periods besides the common ones given above), there exists an algebraic relation of the form*

$$F[f_1(u), f_2(u)] = 0.$$

For example, the following identity holds between the Weierstrass elliptic functions $\wp(u)$ and $\wp'(u)$ of orders 2 and 3 respectively:

$$[\wp'(u)]^2 - 4\wp^3(u) + g_2\wp(u) + g_3 = 0.$$

A natural question arises: could one decrease the number $p+1$ in the general theorem? A negative answer to this question is provided by the following

THEOREM. *If $f(u)$ is a nondegenerate function from the field of Abelian functions $\mathfrak{K}$, then there exist p-vectors $c^{(1)}, \ldots, c^{(p)}$, which do not depend on u, such that p functions of the same field*

$$f_1(u) = f(u + c^{(1)}), \ldots, f_p(u) = f(u + c^{(p)}) \tag{IV.5.2}$$

are not related among themselves by any algebraic (or even by any analytic) relation.

To prove this it suffices to show that the Jacobian of the system of functions (IV.5.2) is not identically zero, i.e.,

$$J(u) = \left|\frac{\partial f_k(u)}{\partial u_j}\right| \not\equiv 0 \tag{IV.5.3}$$

when an appropriate choice of $c^{(j)}$, $j = 1, \ldots, p$, is made. In particular, it is enough to choose $c^{(1)}, \ldots, c^{(p)}$ so that

$$J(0) = \left|\frac{\partial f_k}{\partial u_j}\right| \neq 0, \quad \text{where } \frac{\partial f_k}{\partial u_j} = \left.\frac{\partial f(u + c^{(k)})}{\partial u_j}\right|_{u=0} \quad (k = 1, \ldots, p). \tag{IV.5.3$'$}$$

But the determinant (IV.5.3) coincides with the determinant (II.3.7) which was considered in the proof of Theorem II.3.1. It was established there that if $|\partial f_k/\partial u_j| = 0$ for any choice of the system $c^{(j)}$, $j = 1, \ldots, p$, then the function $f(u)$ must be degenerate. Therefore under our conditions there exist systems $c^{(1)}, \ldots, c^{(p)}$ of p-vectors for which (IV.5.3$'$) holds. The stated theorem follows from this.

By altering the arguments which were used to prove the theorem of Weierstrass and Poincaré one can obtain a stronger result.

THEOREM. *If* $f_1(u), \ldots, f_p(u)$ *is a set of algebraically independent Abelian functions of the field* $\mathfrak{K}$, *at least one of which is nondegenerate, then one more function* $f_0(u)$ *of the same field can be added to the list in such a way that any function* $f(u)$ *of the field can be represented as a rational combination of* $f_0(u), \ldots, f_p(u)$, *i.e.*,

$$f(u) = R[f_0(u), \ldots, f_p(u)].$$

We do not prove this theorem here, but refer the reader to the book of C. L. Siegel mentioned in Chapter I.

§6. Abelian and Picard manifolds

Now let S be the affine t-dimensional (complex) space, the coordinates of whose points we denote by $y_1, y_2, \ldots, y_t$. We denote by S_t the corresponding projective space with homogeneous coordinates $x_0, x_1, \ldots, x_t$ (where $x_0, x_1, \ldots x_t$ cannot be equal to zero simultaneously). From the field $\mathfrak{K}$ of Abelian functions corresponding to a Riemann matrix Ω, we choose some t Abelian functions and represent them as quotients, with common denominator, of the corresponding Jacobian functions:

$$f_1(u) = \frac{\varphi_1(u)}{\varphi_0(u)}, \ldots, f_t(u) = \frac{\varphi_t(u)}{\varphi_0(u)}.$$

Now we set

$$y_1 = f_1(u), \ldots, y_t = f_t(u) \tag{IV.6.1}$$

and let u run through the points of the fundamental parallelotope of periods

$$\Pi_0 \colon u = \alpha_1 \omega^{(1)} + \cdots + \alpha_{2p} \omega^{2p} = \Omega \begin{pmatrix} \alpha_1 \\ \vdots \\ \alpha_{2p} \end{pmatrix},$$

$$0 \leq \alpha_k < 1, \qquad k = 1, \ldots, 2p.$$

Recall (section II.5) that Π_0 is a bounded subset of a p-dimensional complex space $\mathbf{C}^p$ of real dimension $2p$, such that for each point $u \in \mathbf{C}^p$ there is a unique point $u^{(0)}$ in Π_0 *congruent* to it, and this point is related to u by a relation of the type

$$u = u^{(0)} + m_1 \omega^{(1)} + \cdots + m_{2p} \omega^{(2p)} \qquad (m_1, \ldots, m_p \text{ are integers}).$$

Hence any values taken by the functions $f_1(u), \ldots, f_t(u)$ over the entire space $\mathbf{C}^p$ are already taken over the points of the fundamental parallelotope. From now on we will use, instead of the parallelotope Π_0, the p-dimensional torus T_p corresponding to it (see the end of the section II.5). Abelian functions can be considered as meromorphic functions on the torus T_p (i.e., as functions representable as a quotient of two analytic functions in a neighborhood of each point). Correspondingly, the set of points of the affine space S whose coordinates are expressed by the formulas (IV.6.1) forms an image of the torus W.

However, the functions (IV.6.1) are defined neither at poles nor at the points of indeterminacy of the meromorphic functions $f_j(u)$. By using the homogeneous coordinates

$$x_0 = \varphi_0(u), x_1 = \varphi_1(u), \dots, x_t = \varphi_t(u) \tag{IV.6.2}$$

of the corresponding projective space S_t, we improve the situation somewhat in the sense that, to the points of the torus at which $\varphi_0(u)$ becomes zero but at least one of the functions $\varphi_1(u), \dots, \varphi_t(u)$ is not zero, there correspond points of the set W, provided that W is included in the projective space S_t mentioned above.

However, no points of the projective space S_t correspond in general to the points at which all the functions $\varphi_0(u), \dots, \varphi_t(u)$ become zero simultaneously. These exceptional points obviously form a closed subset of the torus T_p. This subset has dimension no greater than $p-1$ since the set of points at which $\varphi_0(u) = 0$ has dimension $p-1$.

The manifold W of the projective space S_t such that W admits a parametric representation of the form (IV.6.2) on the torus T_p (with the exception, mentioned above, of the possible points of indeterminacy) is called *an Abelian manifold corresponding to the given field of Abelian functions.*

Let now $t = p$ and let the functions $f_1(u), \dots, f_p(u)$ from the field $\mathfrak{K}$ be algebraically independent (see section IV.5). We represent them again as quotients of Jacobian functions of the same type with common denominator $\varphi_0(u)$ and adjoin to their list one more arbitrary function $\varphi_{p+1}(u)$ of the same type. We get a parametric representation of a certain Abelian manifold

$$x_0 = \varphi_0(u), x_1 = \varphi_1(u), \dots, x_{p+1} = \varphi_{p+1}(u). \tag{IV.6.3}$$

Among the first $p+1$ functions (IV.6.3) there can be no algebraic relation because otherwise it would be homogeneous (as a consequence of the properties of the generalized periodicity of Jacobian functions of the same type) and would imply an algebraic relation among p Abelian functions $f_j(u)$ $(j = 1, \dots, p)$. However, an algebraic (homogeneous) relation can exist among all the Jacobian functions (IV.6.3). In other words, for some irreducible homogeneous polynomial $\Phi(x_0, \dots, x_{p+1})$ we have an identity

$$\Phi[\varphi_0(x), \dots, \varphi_p(u), \varphi_{p+1}(u)] = 0. \tag{IV.6.4}$$

The polynomial Φ is defined uniquely up to a constant factor. Indeed, if we assumed the existence of another such polynomial, we would be able to eliminate $\varphi_{p+1}(u)$ from the two equations $\Phi = 0$ and $\Phi_1 = 0$ and thus get an algebraic relation among $\varphi_0(u), \dots, \varphi_p(u)$, which is impossible.

Thus we can assert that W is a subset of an algebraic hypersurface of the projective space S_{p+1} where the hypersurface is defined by an irreducible equation

$$\Phi(x_0, \dots, x_{p+1}) = 0.$$

On the other hand the Jacobian

$$J = \begin{vmatrix} \frac{\partial f_1}{\partial u_1} & \cdots & \frac{\partial f_1}{\partial u_p} \\ \cdots & \cdots & \cdots \\ \frac{\partial f_p}{\partial u_1} & \cdots & \frac{\partial f_p}{\partial u_p} \end{vmatrix}$$

is not equal to zero because of our previous assumptions. Therefore it is nonzero everywhere in a neighborhood of some point of the torus. Hence the set of points where the Jacobian is equal to zero is at most $(p-1)$-dimensional. By excluding from T_p the union (denoted by Γ) of two sets: the set of zeros of $\varphi_0(u)$ and the set of zeros of $J(u)$, we can guarantee that, at the remaining points of the torus, the mapping (IV.6.1) is defined and locally invertible. Thus W, which is the image of T_p minus Γ, is a p-dimensional manifold contained in an algebraic hypersurface $\Psi = 0$.

It is useful to express the condition $J \neq 0$ in terms of the functions $\varphi_j(u)$, $j = 0, \dots, p$. Assume that $\varphi_0(u) \neq 0$ at a given point $u \in T_p$. We obtain

$$J = \frac{1}{\varphi_0^{2p}} \begin{vmatrix} \frac{\partial \varphi_1}{\partial u_1}\varphi - \frac{\partial \varphi_0}{\partial u_1}\varphi_1 & \cdots & \frac{\partial \varphi_p}{\partial u_1}\varphi_0 - \frac{\partial \varphi_0}{\partial u_1}\varphi_p \\ \cdots & \cdots & \cdots \\ \frac{\partial \varphi_1}{\partial u_p}\varphi_0 - \frac{\partial \varphi_0}{\partial u_p}\varphi_1 & \cdots & \frac{\partial \varphi_p}{\partial u_p}\varphi_0 - \frac{\partial \varphi_0}{\partial u_p}\varphi_p \end{vmatrix}$$

$$= \frac{1}{\varphi_0^{2p}} \begin{vmatrix} 1 & 0 & \cdots & 0 \\ \frac{\partial \varphi_0}{\partial u_1} & \frac{\partial \varphi_1}{\partial u_1}\varphi_0 - \frac{\partial \varphi_0}{\partial u_1}\varphi_1 & \cdots & \frac{\partial \varphi_p}{\partial u_1}\varphi_0 - \frac{\partial \varphi_0}{\partial u_1}\varphi_p \\ \cdots & \cdots & \cdots & \cdots \\ \frac{\partial \varphi_0}{\partial u_p} & \frac{\partial \varphi_1}{\partial u_p}\varphi_0 - \frac{\partial \varphi_0}{\partial u_p}\varphi_1 & \cdots & \frac{\partial \varphi_p}{\partial u_p}\varphi_0 - \frac{\partial \varphi_0}{\partial u_p}\varphi_p \end{vmatrix}$$

$$= \frac{1}{\varphi_0^{p+1}} \begin{vmatrix} \varphi_0 & \varphi_1 & \cdots & \varphi_p \\ \frac{\partial \varphi_0}{\partial u_1} & \frac{\partial \varphi_1}{\partial u_1} & \cdots & \frac{\partial \varphi_p}{\partial u_1} \\ \cdots & \cdots & \cdots & \cdots \\ \frac{\partial \varphi_0}{\partial u_p} & \frac{\partial \varphi_1}{\partial u_p} & \cdots & \frac{\partial \varphi_p}{\partial u_p} \end{vmatrix}.$$

Therefore the condition $J(u) \neq 0$ is equivalent to $J^*(u) \neq 0$ (when $\varphi_0(u) \neq 0$), where

$$J^*(u) = \begin{vmatrix} \varphi_0 & \varphi_1 & \cdots & \varphi_p \\ \frac{\partial \varphi_0}{\partial u_1} & \frac{\partial \varphi_1}{\partial u_1} & \cdots & \frac{\partial \varphi_p}{\partial u_1} \\ \cdots & \cdots & \cdots & \cdots \\ \frac{\partial \varphi_0}{\partial u_p} & \frac{\partial \varphi_1}{\partial u_p} & \cdots & \frac{\partial \varphi_p}{\partial u_p} \end{vmatrix}.$$

Thus among the Abelian manifolds corresponding to a given Riemann matrix Ω there always exist p-dimensional ones. Let such a manifold W_p be represented parametrically in the form (IV.6.2)

$$x_0 = \varphi_0(u),\ x_1 = \varphi_1(u),\ \dots,\ x_t = \varphi_t(u),$$

where $\varphi_j(u)$, $j = 0, \dots, t$, are Jacobian functions of the same type corresponding to Ω.

It can be proved that, *by adjoining to* W_p *all its limit points, we span the algebraic surface containing* W_p (Conforto, pp. 146–148).

We also state a fundamental theorem which establishes the existence of homeomorphic images of the torus T_p among the Abelian manifolds corresponding to any field $\mathfrak{K}$.

THEOREM OF LEFSCHETZ. *For each Abelian field* $\mathfrak{K}$ *there exist Abelian manifolds, which represent nonsingular algebraic surfaces in the complex projective space* S_t, *for which the corresponding mapping*

$$x_0 = \varphi_0(u), \ldots, x_t = \varphi_t(u)$$

is well defined and homeomorphic on the entire torus T_p. *Moreover, we can also require all functions of the field* $\mathfrak{K}$ *to be rational functions of the Abelian functions* $\varphi_j(u)/\varphi_0(u)$, $j = 1, \ldots, t$.

Compare the last condition with the theorem stated at the end of section IV.5. The proof can be found in the Russian edition of the book by Siegel mentioned on page viii (pp. 162–165) or in the book by Conforto (pp. 158–164).

Abelian manifolds satisfying the conditions of the theorem of Lefschetz are called *Picard manifolds* (or *varieties*).

§7. Rational functions on Picard varieties. Picard integrals of the first kind[24]

Let V_p be a Picard manifold with the parametric representation

$$y_1 = f_1(u), y_2 = f_2(u), \ldots, y_t = f_t(u), \tag{IV.7.1}$$

where the Abelian functions $f_j(u)$ $(j = 1, \ldots, t)$ belong to the field $\mathfrak{K}$ generated by a Riemann matrix Ω. By the definition of a Picard manifold, each function $f(u)$ of the field $\mathfrak{K}$ is a rational function of the coordinates $(y_1, \ldots, y_t)$ of a point P that runs through V_p. Of course the converse is also true: each rational function of the coordinates $(y_1, \ldots, y_t)$ of the point $P \in V_p$ belongs to $\mathfrak{K}$.

We make the following definition.

DEFINITION. Two algebraic varieties are called *birationally equivalent* if each is the image of the other under a rational transformation.

We can now state the following theorem.

THEOREM IV.7.1. *All Picard varieties corresponding to the same field* $\mathfrak{K}$ *of Abelian functions are birationally equivalent, and conversely each Picard variety birationally equivalent to a given one, denoted by* V_p, *is a Picard variety corresponding to the same field* $\mathfrak{K}$.

PROOF. If $V_p(y_1, \ldots, y_t)$ and $V'_p(y'_p, \ldots, y'_\tau)$ are two Picard varieties corresponding to the same field $\mathfrak{K}$ of Abelian functions, then the coordinates

[24] A more thorough treatment of the results given here can be found by the reader in the monograph by Conforto (pp. 164–174).

$y'_1, \dots, y'_\tau$ as well as $y_1, \dots, y_t$ are functions in the same field $\mathfrak{K}$ (i.e. they are Abelian functions). According to what we said above, y'_j $(j = 1, \dots, \tau)$ are rational functions of $y_1, \dots, y_t$ and y_k $(k = 1, \dots, t)$ are rational functions of $y'_1, \dots, y'_\tau$:

$$y'_j = R'_j(y_1, \dots, y_t), \qquad j = 1, \dots, \tau$$

and

$$y_k = R_k(y'_1, \dots, y'_\tau), \qquad k = 1, \dots, t \tag{IV.7.2}$$

(here R'_j and R_k are rational functions on V'_p and V_p). Thus the birational equivalence of V_p and V'_p is established (moreover, the points $(y'_1, \dots, y'_\tau)$ and $(y_1, \dots, y_t)$ must always correspond to the same point of the torus T_p).

Now let V'_p be an arbitrary variety birationally equivalent to V_p. For each $u \in T_p$ the coordinates $(y_1, \dots, y_t)$ of the point $P \in V_p$ are Abelian functions of u. Hence the coordinates $(y'_1, \dots, y'_\tau)$ of the point $P' \in V'_p$ are also Abelian functions on the same field $\mathfrak{K}$. Therefore V'_p is an Abelian manifold. Moreover, by the second group of equations (IV.7.2), a unique point $P \in V_p$ and therefore a unique point $u \in T_p$, corresponds to each point $P' \in V'_p$. Hence V'_p (like V_p) is a homeomorphic image of the torus T_p. Finally, each function $f(u) \subset \mathfrak{K}$ is a rational function of the coordinates of the point $P \in V_p$, and by (IV.7.2) it is a rational function of the coordinates of the point $P' \in V'_p$. We have proved that V'_p is a Picard variety corresponding to the field.

We now show that each Abelian manifold can be obtained by means of a rational transformation of a Picard variety. The following theorem summarizes the earlier results about Abelian manifolds.

THEOREM IV.7.2. *All Abelian manifolds corresponding to the field of Abelian functions $\mathfrak{K}$ generated by some Riemann matrix Ω are algebraic varieties (i.e., each of them can be defined by a finite number of algebraic equations). Each of them can be obtained by a rational transformation of a Picard variety and conversely an image of a rational transformation of a Picard variety (or an Abelian manifold) is an Abelian manifold (which we can now call an Abelian variety) of the field $\mathfrak{K}$.*

PROOF. Let V_p be a Picard variety corresponding to the field $\mathfrak{K}$ and let W be any Abelian manifold corresponding to the same field $\mathfrak{K}$. All the coordinates of a point $P \in W$ are rational functions of a point on the Picard variety V_p and therefore W is obtained from V_p by means of a rational transformation, which in general is not rationally invertible (birational). Since V_p is an algebraic variety, it follows from this that W is also an algebraic variety.

Conversely, if W' is the image of W (which is an Abelian manifold or, in particular, a Picard variety) under some rational transformation, then a parametric representation of W' is expressed by rational functions of the coordinates of a variable point in W. But these coordinates are Abelian functions from the field $\mathfrak{K}$. Therefore a parametric representation of W' is also obtained by means of Abelian functions of the same field, and W' is an Abelian manifold.

We return again to the Picard variety V_p, which is a p-dimensional subvariety of a projective space S_t.

By means of a *regular* projection (stereographic projection whose center lies outside V_p and each of whose projecting rays meets V_p at only one point) we can (if $t > p+1$) map V_p one-to-one onto a hypersurface V_p'.

Since this process uses only rational operations on the coordinates and since the mapping is one-to-one and onto, V_p' is again a Picard manifold (its projective model).

In nonhomogeneous coordinates the equation of V_p' takes the form

$$F(y_1, \dots, y_{p+1}) = 0, \tag{IV.7.3}$$

and it becomes an identity if we set

$$y_1 = f_1(u), \dots, y_{p+1} = f_{p+1}(u). \tag{IV.7.4}$$

Without loss of generality, we can assume that the first p of these functions are algebraically independent, so that the Jacobian is not zero:

$$J = \left| \frac{\partial f_j(u)}{\partial u_k} \right| \neq 0, \qquad j, k = 1, \dots, p. \tag{IV.7.5}$$

Expressing the total differentials of the coordinates $y_1, \dots, y_p$ in terms of the differentials u_k, we get

$$dy_j = \sum_{k=1}^{p} \frac{\partial f_j(u)}{\partial u_k} du_k \qquad (j = 1, \dots, p). \tag{IV.7.6}$$

By (IV.7.5) these equations are solvable with respect to $du_1, \dots, du_p$. We notice that $\partial f_j(u)/\partial u_k$ are Abelian functions in $\mathfrak{K}$ and therefore are rational functions on V_p'. Hence, when we solve the system (IV.7.6) in terms of du_k we again obtain rational functions of a variable point on V_p' as the coefficients of dy_j, i.e.,

$$du_k = \sum_{j=1}^{p} q_{kj}(y)\, dy_j \qquad (k = 1, \dots, p). \tag{IV.7.7}$$

We emphasize that $q_{kj}(y)$ are functions of the $p+1$ coordinates of a point of the variety V_p':

$$y_1 = f_1(u), \dots, y_{p+1} = f_{p+1}(u).$$

As a consequence of (IV.7.7), to each point $u \in \mathbf{C}^p$ there corresponds, in general, a unique point P of the hypersurface V_p' (exceptional points of the space $\mathbf{C}^p$ are the points at which all the functions (IV.7.4) become zero).

The converse is not true: the preimages of P in $\mathbf{C}^p$ are all the points of $\mathbf{C}^p$ congruent to u, i.e., differing from u by any period. Hence the functions u_k, $k = 1, \ldots, p$, defined by the system of equations (IV.7.6), are infinite-valued functions of a variable point P on the Picard variety V_p'. Our immediate task is to study these functions.

If to the point $(0, \ldots, 0) \in \mathbf{C}^p$ there corresponds $P_0(y_{1,0}, \ldots, y_{p+1,0})$, a point on V_p', then by integrating the system (IV.7.7) we get

$$u_k = \int_{P_0}^{P} \sum_{j=1}^{p} q_{kj}(y)\, dy_j, \qquad k = 1, \ldots, p, \tag{IV.7.8}$$

where the integration is taken over paths connecting P_0 with P on V_p'.

These integrals are *algebraic*. Indeed, even though $q_{kj}(y)$ are rational functions of the coordinates $y_1, \ldots, y_p, y_{p+1}$ (i.e., Abelian functions of $u_1, \ldots, u_p$), y_{p+1} is related to $y_1, \ldots, y_p$ by an algebraic equation (this is a consequence of the Weierstrass-Poincaré theorem). Hence $q_{kj}(y)$ is an algebraic function of the coordinates $y_1, \ldots, y_p$ on V_p'. Furthermore, even though $u_1, \ldots, u_p$ are infinitely multiple-valued functions of the coordinates $y_1, \ldots, y_{p+1}$ of the point P, they take only a discrete set of values at p and hence the values of the integrals cannot change if the paths of integration connecting P_0 with P are varied continuously in V_p', i.e., changed in such a way that one can be continuously deformed into another without leaving V_p' in the process. Hence these integrals are *integrals of total differentials* and must satisfy the standard conditions of integrability

$$\frac{dq_{k\sigma}}{dy_s} = \frac{dq_{ks}}{dy_\sigma}, \qquad s, \sigma = 1, \ldots, p. \tag{IV.7.9}$$

Here d/dy_s and d/dy_σ are "total" derivatives. To compute them, we must take into account that $q_{k\sigma}$ and q_{ks} depend on $y_1, \ldots, y_p$ not only explicitly but also by means of an algebraic function y_{p+1} of these coordinates. We note finally that all algebraic integrals on the right-hand sides of (IV.7.8) remain finite on V_p'. By analogy with Abelian integrals of the first kind they are called *Picard integrals of the first kind on the Picard variety* V_p', in honor of E. Picard who first introduced them in the case $p = 2$ in 1885.

THEOREM IV.7.3. *The inversion of the transformation formulas* (IV.7.4) *which map p-dimensional complex space* $\mathbf{C}^p$ *onto Picard varieties* V_p' *(which is also a solution of the system of differential equations* (IV.7.6)) *can be accomplished by means of* p *linearly independent Picard integrals of the first kind on the variety* V_p' *according to* (IV.7.8).

We need only prove the linear independence of the Picard integrals. However, this follows immediately from the linear independence of the complex variables $u_1, \ldots, u_p$.

THEOREM IV.7.4. *On a p-dimensional Picard variety V'_p which corresponds to the field $\mathfrak{K}$ of Abelian functions of p variables $u_1, \dots, u_p$, there exist exactly p linearly independent integrals of the first kind which, together with the identity, form a linear space of $p+1$ dimensions.*

PROOF. Indeed, let

$$v = \int_{P_0}^{P} \sum_{j=1}^{p} Q_j(y)\, dy_j + \lambda_0 \tag{IV.7.10}$$

be a Picard integral of the first kind on V'_p. Then by using (IV.7.4) and (IV.7.6) we can represent it in the form of an integral with respect to the variables u:

$$v = \int_0^u \sum_{j=1}^{p} g_j(u)\, du_j + \lambda_0 . \tag{IV.7.11}$$

This integral is extended over the paths in $\mathbf{C}^p$ which connect the points 0 $(0, \dots, 0)$ with $u(u_1, \dots, u_p)$ and which are, as a consequence of (IV.7.4), preimages of the path connecting P_0 with P on V'_p. But $Q_j(y)$ are rational functions of the coordinates $y_1, \dots, y_{p+1}$ of a point on the Picard variety V'_p and therefore $Q_j(y)$ are Abelian functions in the field $\mathfrak{K}$. Since $g_j(u)$ are rational combinations of these functions and the functions $q_{ki}(y)$ (from (IV.7.7)), which are themselves Abelian functions of the same field $\mathfrak{K}$, the functions $g_j(u)$ in (IV.7.11) are also Abelian functions. They cannot have singular points at the finite points of $\mathbf{C}^p$, since v is finite on any finite path of $\mathbf{C}^p$ by the definition of the Picard integral of the first kind. Hence the meromorphic functions $g_j(u)$ must actually be *entire functions*. Their absolute values over the entire space $\mathbf{C}^p$ have the same upper bound as over the torus T_p. From this it follows by Liouville's theorem (which generalizes without change to entire functions of any number of variables) that all the functions $g_j(u)$ are constants, i.e., $g_j(u) = \lambda_j$ $(j = 1, \dots, p)$. Thus the formula (IV.7.11) yields

$$v = \int_0^u \sum_{j=1}^{p} \lambda_j\, du_j + \lambda_0 = \sum_{j=1}^{p} \lambda_j u_j + \lambda_0 ,$$

which completes the proof of the theorem.

In conclusion, we note that when we pass from the notion of an Abelian integral of the first kind to the notion of a Picard integral of the first kind, it becomes possible to consider the system of equations (IV.7.6) as a generalization of the Jacobi problem (after differentiation of the equations occurring in it). Then (IV.7.4) can be considered as a generalization of the solution of the Jacobi problem: here the upper limits of Picard integrals are expressed as Abelian (general and not special) functions of p complex variables.

Appendix

§A. Skew-Symmetric Determinants

A.1. The Pfaffian. A matrix

$$A = (a_{jk}), \qquad j,\ k = 1, \dots, n, \tag{A.1.1}$$

is called *skew-symmetric* if

$$a_{jk} = -a_{kj} \quad (\text{in particular, } a_{jj} = 0). \tag{A.1.2}$$

The corresponding determinant $|A|$ is also called skew-symmetric.

The equalities (A.1.2) can be written in the form

$$\widetilde{A} = -A. \tag{A.1.2$'$}$$

This means that $|\widetilde{A}| = (-1)^n|A|$ for the determinant $|A|$. On the other hand, $|\widetilde{A}| = |A|$. Hence $|A| = (-1)^n|A|$ and therefore when n is odd we have $|A| = 0$.

Let $n = 2m-1$. Consider the minors M_{jk} and M_{kj} of the determinant. If we transpose M_{kj}, we get a minor of the matrix $\widetilde{A}$ with index j, k. Because of $(A.1.2')$ it follows that $M_{kj} = (-1)^{n-1}M_{jk} = M_{jk}$. Thus

$$M_{kj} = M_{jk}, \qquad j,\ k = 1, \dots, 2m-1. \tag{A.1.3}$$

We denote by A_{jk} the term corresponding to the entry a_{jk} in the expansion for the determinant $|A|$. Then $A_{jk} = (-1)^{j+k}M_{jk}$, and under the same conditions we have

$$A_{jk} = A_{kj}, \qquad j,\ k = 1, \dots, 2m-1. \tag{A.1.4}$$

Since $|A| = 0$, it follows in addition that

$$A_{jj}A_{kk} - A_{jk}^2 = 0, \qquad j,\ k = 1, \dots, 2m-1. \tag{A.1.5}$$

In fact, if the rank of the matrix A is less than $n-1 = 2m-2$, then all $A_{jk} = 0$ and (A.1.5) obviously holds. If, on the other hand, this rank is equal to $n-1$, we conclude, on the basis of the equations $\sum_{k=1}^{n} a_{jk}A_{jm} = 0$, which hold for each $m = 1, \dots, n$, that the n-dimensional vectors with coordinates $A_{1m}, \dots, A_{nm}$ are collinear so that

$$\frac{A_{1k}}{A_{1j}} = \frac{A_{2k}}{A_{2j}} = \cdots = \frac{A_{nk}}{A_{nj}}; \tag{A.1.5$'$}$$

and in particular

$$\frac{A_{jk}}{A_{jj}} = \frac{A_{kk}}{A_{kj}}.$$

By (A.1.4), it follows from this that (A.1.5) is true.

We now turn to the case $n = 2m$ and prove that in this case *there exists a form* $\mathscr{P}$ *of degree* m *in the elements of the determinant of* A, *all of whose coefficients are integers, and with the property that*

$$|A| = \mathscr{P}^2. \tag{A.1.6}$$

This form can be constructed explicitly in terms of the entries of the matrix A. It is called the *Pfaffian* after Johann Friedrich Pfaff (1765–1825), an older contemporary of Gauss. However, we will not give the actual construction of the Pfaffian here and will limit ourselves to proving the theorem stated above.

It is most easily proved by induction. In the case $n = 2$ we have

$$|A| = \begin{vmatrix} 0 & a_{12} \\ -a_{21} & 0 \end{vmatrix} = a_{12}^2$$

and we can set $\mathscr{P} = a_{12}$. Assume that the theorem has been proved for $n = 2m - 2$ $(m \geq 2)$. We consider a determinant B of order $2m - 1$ which is a minor of the entry $a_{2m,2m}$ in $|A|$:

$$|B| = \begin{vmatrix} a_{1,1} & \cdots & a_{1,2m-1} \\ a_{2m-1,1} & \cdots & a_{2m-1,2m-1} \end{vmatrix}.$$

It is obvious that $|B|$ is skew-symmetric simultaneously with $|A|$. Since the principal minors A_{rr} $(r = 1, \dots, 2m - 1)$ of the latter determinant are also skew-symmetric and their order is $2m - 2$, by the inductive assumption we get

$$A_{rr} = \mathscr{P}_r^2, \qquad r = 1, \dots, 2m - 1, \tag{A.1.7}$$

where $\mathscr{P}_r$ is a form of degree $m - 1$ in the corresponding elements a_{ij} with integral coefficients.

Assume that among the forms $\mathscr{P}_r$ there is at least one which is not identically zero. Without loss of generality, we can assume that it is $\mathscr{P}_1$. The equality $A_{11} = \mathscr{P}_1^2$ is satisfied by the two forms $\mathscr{P}_1$ and $-\mathscr{P}_1$. We choose one of them and denote it by $\mathscr{P}_1$. We now choose $\mathscr{P}_2, \dots, \mathscr{P}_{2m-1}$ (at each step we select one of the two forms which differ only by the factor -1) in such a way that we have

$$A_{1r} = \mathscr{P}_1 \cdot \mathscr{P}_r, \qquad r = 1, \dots, 2m - 1,$$

by (A.1.5). Now by (A.1.5′) we have

$$A_{11}A_{rs} - A_{1s}A_{r1} = 0, \qquad \mathscr{P}_1^2 A_{rs} = \mathscr{P}_1^2 \mathscr{P}_r \mathscr{P}_s,$$

and since $\mathscr{P}_1 \not\equiv 0$ we have

$$A_{rs} = \mathscr{P}_r \mathscr{P}_s, \qquad r, s = 1, \ldots, 2m-1. \tag{A.1.8}$$

We notice, that in the case when $\mathscr{P}_1, \ldots, \mathscr{P}_{2m-1}$ are all identically zero, all A_{rr} are also zero and hence all A_{rs} are also zero, as follows from (A.1.5). Therefore the equalities (A.1.8) hold in this case.

Turning now to the determinant $|A|$, we expand it by the entries of the last column and then we expand each of the minors of these elements (except the principal one, which is equal to $|B| = 0$) by the elements of the last row. We get

$$|A| = -\sum_{r,s=1}^{2m-1} a_{r,n} a_{n,s} A_{rs} = \sum_{r,s=1}^{n-1} a_{r,n} a_{s,n} A_{rs},$$

or, by substituting for A_{rs} their expressions (A.1.8),

$$|A| = \sum_{r,s=1}^{n-1} a_{r,n} a_{s,n} \mathscr{P}_r \mathscr{P}_s = (a_{1,n}\mathscr{P}_1 + \cdots + a_{n-1,n}\mathscr{P}_{n-1})^2. \tag{A.1.9}$$

It remains only to set

$$\mathscr{P} = a_{1,n}\mathscr{P}_1 + \cdots + a_{n-1,n}\mathscr{P}_{n-1}. \tag{A.1.10}$$

As an illustration, we consider the case $n = 4$. Here

$$|A| = \begin{vmatrix} 0 & a & b & d \\ -a & 0 & c & e \\ -b & -c & 0 & f \\ -d & -e & f & 0 \end{vmatrix} \quad \text{and} \quad |B| = \begin{vmatrix} 0 & a & b \\ -a & 0 & c \\ -b & -c & 0 \end{vmatrix}.$$

In our case

$$A_{11} = \begin{vmatrix} 0 & c \\ -c & 0 \end{vmatrix} = c^2, \qquad A_{22} = \begin{vmatrix} 0 & b \\ -b & 0 \end{vmatrix} = b^2, \qquad A_{33} = \begin{vmatrix} 0 & a \\ -a & 0 \end{vmatrix} = a^2.$$

We set $\mathscr{P}_1 = c$. Since $A_{12} = bc$ and $A_{13} = ac$, we have $\mathscr{P}_2 = b$ and $\mathscr{P}_3 = a$. Therefore, according to (A.1.9), we have

$$|A| = (dc + eb + fa)^2, \quad \text{i.e.,} \quad \mathscr{P} = cd + be + af.$$

A.2. The Frobenius Theorem. Let P be a skew-symmetric $(2p, 2p)$ matrix, all of whose entries are integers, and with $|P| \neq 0$. For each unimodular $(2p, 2p)$ matrix M with integral entries we can construct a transformed matrix

$$P^* = MP\widetilde{M}. \tag{A.2.1}$$

It is obvious that the entries of the matrix P^* are also integers. Moreover, $\widetilde{P}^* = M\widetilde{P}\widetilde{M} = -MP\widetilde{M} = -P^*$, i.e., P^* is skew-symmetric and (because of (A.2.1)) we have

$$|\widetilde{P}^*| = |M|\,|P|\,|\widetilde{M}| = |P|\,|M|^2 = |P|.$$

It is easy to check that the relation between the skew-symmetric $(2p, 2p)$ matrices P and P^* with integral entries and nonzero determinants, given by an equation of the form (A.2.1), is an *equivalence relation*. In particular, if we start with P and perform, one after another, the transformations of the form (A.2.1) which are determined by the unimodular matrices $M_1, \dots, M_n$, where n is any natural number, we stay in the same equivalence class. It is natural to try to select a matrix of the simplest structure from each class in order to have a representative of the class. This is made possible by the following statement.

THE FROBENIUS THEOREM (1878). *For each skew-symmetric $(2p, 2p)$ matrix P with integral entries and nonzero determinant, there exists a unimodular $(2p, 2p)$ matrix M with integral coefficients for which*

$$P^* = MP\widetilde{M} = \operatorname{Diag}\left\{\beta_1\begin{pmatrix} 0 & 1 \\ -1 & 0\end{pmatrix}, \dots, \beta_p\begin{pmatrix} 0 & 1 \\ -1 & 0\end{pmatrix}\right\}, \tag{A.2.2}$$

where $\beta_1, \dots, \beta_p$ are positive integers such that

$$\beta_1 | \beta_2 | \cdots | \beta_p\,. \tag{A.2.3}$$

PROOF. The transformation we are looking for can be obtained step by step by applying unimodular matrices of a particular form. We show that it suffices to use three kinds of matrices.

(A) The matrix M is derived from the unit matrix E_{2p} by permuting the jth and kth rows (or by permuting the jth and kth columns). In this case the matrix $P^* = MP\widetilde{M}$ differs from P only by having jth and kth rows permuted and at the same time its jth and kth columns permuted.

(B) M is derived from the unit matrix by multiplying the jth diagonal entry of the unit matrix by -1. As a result, P^* will differ from P by the signs of the entries in its jth row and jth column (note that the entry which belongs both to the jth row and jth column is zero).

(C) M is derived from the unit matrix by substituting an integer l for the zero at the intersection of the jth row and kth column. As a result $P^* = MP\widetilde{M}$ is obtained from P by adding to the entries of the jth row the corresponding entries of the kth row multiplied by l and then performing the same operation on the entries of the jth and kth columns. We note that we then get the following entry at the intersection of jth row and jth column of the product $MP\widetilde{M}$:

$$a_{jj} + la_{jk} + la_{kj} + l^2 a_{kk}.$$

But this entry is equal to zero, as it should be in a skew-symmetric matrix, since $a_{jj} = a_{kk} = 0$ and $a_{jk} = -a_{kj}$.

Thus, the equivalence class to which the matrix P belongs contains all matrices which can be obtained from P by combining, in any way, a finite number of the elementary transformations just described. We will show that by performing them in an appropriate order we can arrive at the required result.

Let δ be *the least absolute value of all nonzero entries of all matrices* P which belong to the same equivalence class and let P^* be a matrix containing the entry $\pm\delta$. By using the operations (B) we arrive at the case when the entries equal to δ occur among the entries of P^*. Let δ be at the intersection of the j'th row and k'th column, i.e., $a_{j'k'} = \delta$. We show that the rest of the entries of P^* are integral multiples of δ, i.e., δ is the greatest common divisor of all the entries of P^*.

Indeed, if we assume, for example, that $a_{jk'}$ $(j \neq j')$ is not divisible by δ, then $a_{jk'} = \delta(-l) + \delta'$ where l and δ' are integers and moreover $0 < \delta' < \delta$. By applying an appropriate operation (C) we get an entry δ' in place of the entry $a_{jk'}$ which contradicts the definition of δ as the least possible entry. Thus, first of all, the entries of P^* which are in the same row or in the same column with δ are divisible by it. It follows from this that by applying operations (C) to them again we can make them equal to zero (except the entry δ itself, of course). Assume that *after* this we can find an entry a_{jk} $(j \neq j', k \neq k')$ which is not divisible by δ. Then, by means of an operation (C) we can get a new matrix whose jth row is the sum of the jth and j'th rows of the matrix P^*. Moreover, the former entry a_{jk} will occur in the same row as δ and will have to be divisible by δ. Thus we have again obtained a contradiction.

We now notice that the common divisors of the entries of P are obviously preserved by the operations (A), (B), and (C) and hence the g.c.d. of the entries of P is also preserved by these operations. But the g.c.d. of the entries of the matrix P^* that occur in the Frobenius Theorem is equal to β_1. Thus $\beta_1 = \delta$.

By changing the notation, if necessary, we can assume that the matrix P is, from the very beginning, in the form where at least one of its entries is equal to $\delta = \beta_1$ and all the others in the row and the column to which β_1 belongs are equal to zero. Applying an operation (A) we can make the entry β_1 occur second in the first row. Then $-\beta_1$ will be the first entry on the second row (because of skew-symmetry) and therefore the matrix P must be in the form

$$P = \left(\begin{array}{cc|c} 0 & \beta_1 & \\ & & 0 \\ -\beta_1 & 0 & \\ \hline & 0 & P_1 \end{array}\right). \tag{A.2.4}$$

The matrix denoted by P_1 has the same properties as the matrix P, but its order is the order of P minus two. Since β_1 is the g.c.d of *all* entries of P, each entry of P_1 is divisible by β_1. Hence if we denote by β_2 the g.c.d of all the entries of P_1,

$$\beta_1 | \beta_2.$$

It is clear that those transformations of the matrix P of types (A), (B), and (C) which do not change its first two rows and first two columns play the same role with respect to P_1 as they played with respect to the entire matrix P. Hence, merely by the preceding transformations, we transform P_1 to the form

$$P_1 = \left(\begin{array}{cc|c} 0 & \beta_2 & \\ & & 0 \\ -\beta_2 & 0 & \\ \hline \multicolumn{2}{c|}{0} & P_2 \end{array}\right),$$

and we transform the entire matrix P to the form

$$P = \left(\begin{array}{cccc|c} 0 & \beta_1 & 0 & 0 & \\ -\beta_1 & 0 & 0 & 0 & \\ & & & & 0 \\ 0 & 0 & 0 & \beta_2 & \\ 0 & 0 & -\beta_2 & 0 & \\ \hline \multicolumn{4}{c|}{0} & P_2 \end{array}\right).$$

By repeating these steps p times we get P into the form (A.2.2).

In the main part of this book we have, however, used a somewhat different form for a skew-symmetric matrix. By repeatedly moving the rows and columns of the matrix (A.2.2) (i.e., by applying operations (A) repeatedly) we obtain the following representation of P which lies in the same equivalence class:

$$P = \begin{pmatrix} 0 & B \\ -B & 0 \end{pmatrix}, \tag{A.2.5}$$

where B is the diagonal matrix

$$B = \begin{pmatrix} \beta_1 & 0 & \cdots & 0 \\ 0 & \beta_2 & \cdots & 0 \\ \cdots & \cdots & \cdots & \cdots \\ 0 & 0 & \cdots & \beta_p \end{pmatrix}. \tag{A.2.6}$$

Recall that β_1, $\beta_1 \geq 1$, is the g.c.d of all entries of a given $(2p, 2p)$ matrix P and β_2 is the g.c.d of all entries of a $(2p-2, 2p-2)$ matrix P_1 which satisfies (A.2.4), and so on. We also note that

$$|P| = |B|^2 = \beta_1^2 \beta_2^2 \cdots \beta_p^2. \tag{A.2.7}$$

This consequence of the Frobenius Theorem would allow us to avoid referring, in the main part of the book, to the possibility of representing a skew-symmetric determinant as a square of the corresponding Pfaffian.

§B. Divisors of analytic functions

B.1. General theorems. In section II.1 we have already discussed the relation of divisibility between analytic functions. Here we show how the main properties of divisibility can be derived from more general algebraic theorems. These can be found in B. L. Van der Waerden's *Algebra*.

Since the notion of divisibility of analytic functions involves the function in a neighborhood of a given point $a \in \mathbf{C}^p$, we will, without loss of generality, take $a = 0$ in this section. We denote by $\mathfrak{M}_p$ the set of functions of u which are meromorphic at the origin, i.e., which are representable in some neighborhood of the origin as a quotient of two analytic functions. We denote by $\mathfrak{H}_p$ the set of functions analytic at the origin. $\mathfrak{M}_{p-1}$ and $\mathfrak{H}_{p-1}$ denote analogous sets of functions of $p-1$ variables $(u_2, \dots, u_p) = 'u$.

We form two rings of polynomials in u_1 with coefficients in $\mathfrak{M}_{p-1}$ and $\mathfrak{H}_{p-1}$ respectively: $\mathfrak{M}_{p-1}[u_1]$ and $\mathfrak{H}_{p-1}[u_1]$. It is obvious that the pseudopolynomials defined in section II.1 belong to $\mathfrak{H}_{p-1}[u_1]$.

The ring $\mathfrak{M}_{p-1}[u_1]$ is a *Euclidean ring* since the following relations hold for any two of its nonzero elements f and g:

(1) the degree of the product fg is not less than the degree of f;

(2) the representation $f = gh + r$ exists when either $r = 0$ or the degree of r is less than the degree of g (division algorithm).

It follows from this that the elements f and g have a g.c.d., and it can be represented in the following form by means of consecutive divisions (Euclidian algorithm):

$$(f, g) = \text{g.c.d. } (f, g) = ff_1 + gg_1,$$

where f_1 and g_1 are elements of the same ring.

Furthermore, each nonzero element of $\mathfrak{M}_{p-1}[u_1]$ can be uniquely decomposed (up to unit factors) into a product of prime elements. In particular, it follows from this that Euclid's theorem holds.

If $f|gh$ *and* $(f, g) = 1$, *then* $f|h$.

We note that, according to the general theory, the units of our ring are exactly the elements ε which belong to $\mathfrak{M}_{p-1}[u_1]$ together with ε^{-1}. It follows from this that only a polynomial of a nonzero degree can be a unit of the ring: the set of all units of $\mathfrak{M}_{p-1}[u_1]$ coincides with the set of meromorphic functions in $\mathfrak{M}_{p-1}$ which are not identically zero.

This means that a theory analogous to the theory of divisibility of polynomials in one variable with rational coefficients (in this case all nonzero rational numbers form the set of units) holds for the ring $\mathfrak{M}_{p-1}[u_1]$.

In order to move from the ring $\mathfrak{M}_{p-1}[u_1]$ to the ring $\mathfrak{H}_{p-1}[u_1]$ and then to the larger set $\mathfrak{H}_p$ of functions of p variables which are analytic at the origin, we use induction on the number of variables. Note that in the case of one variable the prime elements are reduced (up to multiplication by units) to a single function $\varphi(u_1) = u_1$ and hence the decomposition of $f(u_1) \in \mathfrak{H}_1$

into prime factors has the form $f(u_1) = u_1^{\nu} g(u_1)$ where $\nu \geq 0$ and $g(0) \neq 0$ (compare with section II.1).

Assume now that the theorem about the uniqueness of decomposition into prime factors and all its consequences is true for $\mathfrak{H}_{p-1}$. Let $\varphi(u_1)$ be a nonzero polynomial in $\mathfrak{H}_{p-1}[u_1]$. It is called *primitive* if the g.c.d of the set of its coefficients is equal to 1 (it is enough that the coefficient of the highest power is a nonzero constant and therefore all pseudopolynomials are primitive). If $f(u_1)$ is any nonzero polynomial in $\mathfrak{H}_{p-1}[u_1]$ and $d = d('u)$ is the g.c.d of its coefficients, then $f(u_1)$ can be represented in the form

$$f(u_1) = d('u)\varphi(u_1), \tag{B.1.1}$$

where $\varphi(u_1)$ is a primitive polynomial.

We have the following general facts.

(I) The product of two primitive polynomials is a primitive polynomial (an analog of the *Gauss lemma*).

(II) If $f(u_1) \in \mathfrak{M}_{p-1}[u_1]$, there exists a primitive polynomial $\varphi(u_1) \in \mathfrak{H}_{p-1}[u_1]$ such that

$$f(u_1) = \frac{d('u)}{g('u)}\varphi(u_1), \qquad g, d \in \mathfrak{H}_{p-1}. \tag{B.1.2}$$

In order to establish this under our conditions it is enough to find a common denominator $g('u)$ of the meromorphic coefficients of the polynomial $f(u_1)$ and then to apply a representation of the form (B.1.1) to the numerators.

(III) The polynomial $\varphi(u_1)$ in the formula (B.1.2) is uniquely determined by $f(u_1)$ up to unit factors from $\mathfrak{H}_{p-1}$. Analogously, $f(u_1)$ is uniquely determined by the primitive polynomial $\varphi(u_1)$ up to unit factors from $\mathfrak{M}_{p-1}$.

(IV) The relation between the polynomials in $\mathfrak{M}_{p-1}[u_1]$ and the primitive polynomials in $\mathfrak{H}_{p-1}[u_1]$ associates with a product of two primitive polynomials $\varphi_1(u_1)$ and $\varphi_2(u_1)$ the product of the polynomials $f_1(u_1)$ and $f_2(u_1)$ corresponding to them (and conversely). Hence if the polynomial $f(u_1)$ is prime in $\mathfrak{M}_{p-1}[u_1]$, then $\varphi(u_1)$ is also prime in $\mathfrak{H}_{p-1}[u_1]$ (and conversely).

The following important corollary follows from the facts listed above: *under the inductive assumption the theorem about the uniqueness of the decomposition into prime factors is valid for the ring* $\mathfrak{H}_{p-1}[u_1]$.

In particular, any pseudopolynomial $P(u_1) \in \mathfrak{H}_{p-1}[u_1]$ decomposes into prime factors in a unique way. It follows from this, in view of the Weierstrass preparation theorem, that *the theorem about uniqueness is valid in* $\mathfrak{H}_p$ *as well, i.e., it is valid for functions of* p *variables that are analytic at the origin.* This completes the construction of the theory of divisibility for analytic functions of several variables.

We will discuss a consequence of this theory. The derivation of this consequence requires the use of the simplest properties of the resultant of two polynomials.

Consider two pseudopolynomials

$$P(u_1) = u_1^n + A_1 u_1^{n-1} + \cdots + A_n,$$
$$Q(u_1) = u_1^m + B_1 u_1^{m-1} + \cdots + B_m.$$

We will not use the restriction on their lower orders (see the definition of a pseudopolynomial in section II.1). The fact that $P(u_1)$ and $Q(u_1)$ are primitive polynomials in $\mathfrak{H}_{p-1}[u_1]$ is sufficient. Eliminating u_1 from P and Q, we obtain their *resultant* d in the form of a determinant whose entries are coefficients of the polynomials P and Q. It is obvious that, under our conditions, d belongs to the set $\mathfrak{H}_{p-1}$. The main property of the resultant states that it is identically equal to zero if and only P and Q are not relatively prime, i.e., they have a common divisor different from a unit of $\mathfrak{H}_{p-1}[u_1]$. Since the polynomials P and Q are primitive, this common divisor cannot be a divisor of their coefficients and therefore is a pseudopolynomial of degree not less than 1.

Hence in the case when P and Q are relatively prime, i.e., when $(P, Q) = 1$, the resultant is a function of the $p-1$ variables $(u_2, \dots, u_p) = {}'u$, which is analytic at the origin and is not identically zero. Then it follows from a simple calculation that there exist polynomials P' and Q' in the ring $\mathfrak{H}_{p-1}[u_1]$ such that the condition $PP' + QQ' = d$ holds. Thus *the condition $(P, Q) = 1$ for pseudopolynomials is equivalent to the existence of polynomials P' and Q' in $\mathfrak{H}_{p-1}[u_1]$ such that*

$$PP' + QQ' = d, \tag{B.1.3}$$

where $d = d('u) \in \mathfrak{H}_{p-1}$ and $d \not\equiv 0$.

B.2. Continuation of the divisibility relation from a point to a region. The notion of a divisor of an analytic function depends on the point $a \in \mathbf{C}^p$ under consideration. If, for example, a function $f(u)$ is a unit at a, i.e., $f(a) \neq 0$, then it may be prime or may decompose into a product of several prime functions at a neighboring point b where $f(b) = 0$. However, of course, in a sufficiently small neighborhood U of a in which $f(u) \neq 0$ it will remain a unit for any point $b \in U$.

Consider two functions $f(u)$ and $g(u)$ that are analytic at a point a. If they have a common divisor $h(u)$ $(h(a) = 0)$, then in any sufficiently small neighborhood of a there is a point b for which $h(b) \neq 0$. Therefore $h(u)$ is not a common (*proper*) divisor of $f(u)$ and $g(u)$ at b. At this point $f(u)$ and $g(u)$ may be relatively prime. However, *the relation $(f, g) = 1$ which holds at some points holds also in a sufficiently small neighborhood of* b. This is the content of the following theorem.

THEOREM B.2.1. *If $f(u)$ and $g(u)$ are two functions that are analytic at a and if $(f, g)_a = 1$, then there is a neighborhood $U(a)$ of a such that the condition $(f, g)_b = 1$ holds for each $b \in U(a)$.*

PROOF. Apply the Weierstrass preparation theorem to the functions f and g at the point a (if necessary, we begin by performing an appropriate linear transformation of variables). We get

$$f = P_a \varphi_a, \qquad g = Q_a \psi_a. \tag{B.2.1}$$

Here

$$P_a = (u_1 - a_1)^k + A_1(u_1 - a_1)^{k-1} + \cdots + A_k,$$
$$Q_a = (u_1 - a_1)^l + B_1(u_1 - a_1)^{l-1} + \cdots + B_l,$$

and moreover in some neighborhood $U(a)$ of the point a the functions A_i, B_j are analytic functions of $p-1$ variables $u_2, \ldots, u_p$ and φ_a, ψ_a are analytic functions of p variables $u_1, \ldots, u_p$ such that $\varphi_a(u) \neq 0$ and $\psi_a(u) \neq 0$, i.e., φ_a and ψ_a are units at each point $u \in U(a)$.

Because $(f, g)_a = 1$ we have $(P_a, Q_a) = 1$ and therefore (see B.1.3) there exist functions P'_a, Q'_a, and d_a analytic in the same neighborhood $U(a)$ and such that

$$P_a P'_a + Q_a Q'_a = d_a('u). \tag{B.2.2}$$

Let b be any point in $U(a)$. Assume, contrary to the conclusion of the theorem, that f and g are relatively prime at this point. Then there exist functions $\delta_b(u)$, $\alpha_b(u)$, and $\beta_b(u)$ analytic at the point b and such that

$$f(u) = \delta_b(u)\alpha_b(u), \qquad g(u) = \delta_b(u)\beta_b(u), \tag{B.2.3}$$

where moreover $\delta_b(b) \neq 0$.

From (B.2.1) and (B.2.3) we derive analytic representations for the pseudopolynomials $P_a(u)$ and $Q_a(u)$ such that the same representations can also be used in a neighborhood of a point b:

$$P_a(u) = \varphi_a^{-1}(u)f(u) = \varphi_a^{-1}\alpha_i\delta_b, \qquad Q_a(u) = \psi_a^{-1}\beta_a\delta_b. \tag{B.2.4}$$

Hence the pseudopolynomials $P_a(u)$ and $Q_a(u)$ (like $f(u)$ and $g(u)$) must have a common divisor δ_b at b. But it follows from (B.2.2) that each common divisor of $P_a(u)$ and $Q_a(u)$ at b must be a divisor of the function $d_a('u)$ at the same point, i.e., the common divisor mentioned above does not depend on u_1. However, the existence of such a common divisor for the pseudopolynomials which become zero at b contradicts the fact that they remain primitive at b. Hence theorem (B.2.1) is proved.

The following important corollary follows from this theorem: *the fraction $f(u)/g(u)$, where $f(u)$ and $g(u)$ are analytic at a, remains irreducible in some neighborhood of the point a if it is irreducible at the point a itself.*

We will prove another theorem of a similar character, which we used in section IV.3.

THEOREM B.2.2. *If $\varphi(u) \not\equiv 0$ is a Jacobian function which corresponds to some Riemann matrix Ω, then there exist points $c \in \mathbf{C}^p$ such that the* *following condition holds for the functions* $\varphi(u)$ *and* $\varphi(u+c)$ *at all points* $u \in \mathbf{C}^p$:

$$(\varphi(u), \varphi(u+c)) = 1.$$

PROOF. Let us verify that it is enough to prove the theorem for points which belong to the parallelotope Π_0 of periods determined by the columns of Ω. Indeed, let $(\varphi(u), \varphi(u+c)) = 1$ for each $u \in \Pi_0$. If $u_0^* \notin \Pi_0$, then there are a period ω and a point $u_0 \in \Pi_0$ such that $u_0^* = u_0 + \omega$. By the definition of a Jacobian function,

$$\begin{aligned} \varphi(u+\omega) &= \exp[l(u)+b]\varphi(u), \\ \varphi(u+c+\omega) &= \exp[l(u+c)+b]\varphi(u+c), \end{aligned}$$

where $l(u)$ is a linear function and b is a constant p-vector. It follows that the values $\varphi(u)$ and $\varphi(u+c)$ reproduce, in a neighborhood of u_0^*, the values of the same functions up to unit (exponential) factors in a neighborhood of u_0. Hence there is no common divisor at u_0^* if there is none at u_0.

Let $a = (a_1, \dots, a_p) \in \Pi_0$. We introduce two types of local coordinates $v = u - a$ and $z = \mathscr{L}_a(v)$ in a neighborhood of a. Here $\mathscr{L}_a$ is a linear transformation needed for an application of the Weierstrass preparation theorem. As a result of this transformation, the terms of lowest order occuring in the expansion of the function

$$\varphi^*(z) = \varphi[a + \mathscr{L}_a^{-1}(z)] = \varphi(a+\nu) = \varphi(u)$$

in a power series in a neighborhood of the origin must contain a term of the form z_1^k, where $k = k(a)$ is the degree of the homogeneous polynomial formed by these terms. Then, according to the Weierstrass theorem, we have the representation

$$\varphi(u) = \varphi(a+v) = P_a(v)f_a(v) = P_a^*(z)f_a^*(z), \tag{B.2.5}$$

where

$$P_a^*(z) = z_1^k + A_1('z)z_1^{k-1} + \cdots + A_k('z) = P_a(v), \tag{B.2.6}$$

and moreover $A_j('z)$ $(j = 1, \dots, k)$ are functions analytic at the point $z = 0$ and independent of z_1, and $f_a^*(z)$ is a unit at $z = 0$ (i.e., $f_a^*(0) \neq 0$).

Let $P_a(v)$ and $f_a(v)$ remain analytic at all points of a neighborhood $W(a) : ||v|| < r_a$ and moreover let $f_a(v)$ be nonzero at these points. Denote by $U(a)$ a neighborhood of the same point with half the radius of the preceding one: $||v|| < (1/2)r_a$. Thus with each point $a \in \Pi_0$ we associate its neighborhoods $W(a)$, $U(a)$ and the linear transformation $z = \mathscr{L}_a(v) = \mathscr{L}_a(u-a)$ as well as a representation (B.2.5), where all the functions occurring in it are analytic at $W(a)$.

We select a set of U_a forming a finite cover of the parallelotope Π_0 and let the cover be composed of the neighborhoods $U(a^j)$, $j = 1, \dots, n$, which for brevity we denote by U_j. The corresponding pseudopolynomials $P_{a^{(j)}}(v) = P_{a^{(j)}}^*(z)$ will be distinguished just by their indices j without mentioning the point $a^{(j)}$: $P_j(v) = P_j^*(z)$, $j = 1, \dots, n$. They are all analytic in the same hyperball

$$W\colon \|v\| < r = \min\{r_{a^{(1)}}, \dots, r_{a^{(m)}}\}.$$

Since no P_j is identically zero, one can find points c in any neighborhood of the origin such that all these pseudopolynomials are nonzero at these points. To do this it suffices to first take a point $c^{(1)}$ such that $P_1(c^{(1)}) \neq 0$. Then, we take another point $c^{(2)}$ in a neighborhood V_1 of the first, such that V_1 is small enough to be in W ($V_1 \subset W$) and such that $P_1(v) \neq 0$ for all $v \in V_1$. The second point is chosen so that $P_2(c^{(2)}) \neq 0$ and then a neighborhood $V_2 \subset V_1$ of it is chosen so that $P_2(v) \neq 0$ for all $v \in V_2$. Continuing this process by choosing a third point, and so on, we get a neighborhood V_n of some point $c^{(n)} = c \in \Pi_0$ at which none of the functions $P_1(v), \dots, P_n(v)$ is zero. We put one more restriction on c: $\|c\| < r/2$. Such a point satisfies the condition of the theorem, as we now prove.

For any point $b \in \Pi_0$, there is a neighborhood U_ν $(1 \le \nu \le n)$ among those considered above such that $b \in U_\nu$. We fix this neighborhood. Then the point $u = b + c$ will lie in a larger neighborhood W_ν with the same center, since $v = u - a^{(\nu)}$ for this point and

$$\|v\| = \|(b + c) - a^{(\nu)}\| \le \|b - a^{(\nu)}\| + \|c\| < \frac{1}{2}r + \frac{1}{2}r = r.$$

Hence one can use the same representation of the form (B.2.5) for both of the functions $\varphi(u)$ and $\varphi(u+c)$. This gives us the right to drop the indices which show the point at which a representation is formed and to which the linear transformation of local coordinates $z = \mathscr{L}(v)$ corresponds.

For $u \in U_\nu$ we have

$$\varphi(u) = \varphi(a + v) = P(v)f(v) = P^*(z)f^*(z), \tag{B.2.5$'$}$$

$$\begin{aligned}\varphi(u + c) &= \varphi(a + v + c) = P(v + c)f(v + c) \\ &= P^*(z + \gamma)f^*(z + \gamma),\end{aligned} \tag{B.2.5$''$}$$

where $\gamma = \mathscr{L}(c)$. Moreover,

$$P^*(z) = z_1^k + A_1 z_1^{k-1} + \cdots + A_k = P(v) \tag{B.2.6$'$}$$

is a primitive pseudopolynomial of degree k. It is obvious that

$$P^*(z + \gamma) = z_1^k + B_1 z_1^{k-1} + \cdots + B_k = P(v + c) \tag{B.2.6$''$}$$

is also a primitive pseudopolynomial (it is obtained by substituting $z_1 + \gamma_1, \dots, z_p + \gamma_p$ for $z_1, \dots, z_p$ in $(B.2.6')$ and then grouping the terms with the same power of z_1). We have $z = \mathscr{L}(v) = 0$ when $v = 0$, from which it follows that

$$P^*(0) = P(0) \quad \text{and} \quad P^*(\gamma) = P(c) \neq 0;$$

and therefore the pseudopolynomials $P^*(z)$ and $P^*(z+\gamma)$ are relatively prime at the point $z = 0$. It follows from this that there are a function $d('z)$ (which does not depend on z_1) and pseudopolynomials $R(z)$ and $S(z)$ such that the following identities of the form (B.1.3) hold:

$$R(z)P^*(z) + S(z)P^*(z+\gamma) = d('z). \tag{B.2.7}$$

Since $R(z)$, $S(z)$, and $d('z)$ are obtained from $P^*(z)$ and $P^*(z+\gamma)$ by entire rational operations (see the end of the section B.1), they are (like $P^*(z) = P(v)$ and $P^*(z+\gamma) = P(v+c)$) analytic at all points of the neighborhood of the origin that is the image of the hyperball W under $z = \mathscr{L}(v)$.

Assume that, contrary to the statement of the theorem, $\varphi(u)$ and $\varphi(u+c)$ have a common (proper) divisor at a point $b \in U_\nu$. Since these functions differ from the corresponding pseudopolynomials $P^*(z) = P(v)$ and $P^*(z+\gamma) = P(v+c)$ only by factors which remain units over the entire neighborhood U_ν, it follows from the last assumption that the pseudopolynomials $P^*(z)$ and $P^*(z+\gamma)$ must have a common (proper) divisor $\delta(z)$ at the point β when β corresponds to the point b under the transformation $z = \mathscr{L}(v) = \mathscr{L}(u-a)$, i.e., when $\beta = \mathscr{L}(b-a)$. But it follows from (B.2.7) that $\delta(z)$ is also a divisor of $d('z)$, i.e., that $\delta(z)$ *does not depend on* z_1. We have arrived at a contradiction since the primitive polynomials $P^*(z)$ and $P^*(z+\gamma)$ cannot have a common divisor $\delta(z)$ which does not depend on z_1 and at the same time is not a constant.

B.3. Poincaré-Cousin theorem. It is well known that a meromorphic function of one complex variable can be defined in two ways. It is, on the one hand, a function which does not have any singularities in the finite plane except poles; and on the other hand, a function which can be expressed as a quotient of two entire functions. The equivalence of these two definitions was established by Weierstrass in 1876. The corresponding assertion for two complex variables was first stated and proved by Poincaré in 1883.

THEOREM OF POINCARÉ. *If a function $f(u_1, u_2)$ of two complex variables can be represented as a quotient of two functions $\mathscr{M}_a(u_1, u_2)$ and $\mathscr{N}_a(u_1, u_2)$ in a neighborhood of each point (a_1, a_2), where a_1 and a_2 are any complex numbers and the functions $\mathscr{M}_a(u_1, u_2)$, $\mathscr{N}_a(u_1, u_2)$ are analytic at the point, there exist two entire functions $G_1(u_1, u_2)$ and $G_2(u_1, u_2)$ such that $f(u_1, u_2)$ is their quotient for any u_1 and u_2.*

Of course this theorem is quite analogous to the theorem of Weierstrass. Indeed, if a function of one variable $f(u)$ can be represented in the form of a quotient $\mathscr{M}_a(u)/\mathscr{N}_a(u)$ in a neighborhood of a and if $\mathscr{M}_a(u)$ and $\mathscr{N}_a(u)$ are analytic at a ($\mathscr{N}_a(u) \not\equiv 0$), then a is either a regular point or a pole of $f(u)$. The converse is also true. Hence in the case of one variable the question is resolved in a simple manner. It is enough to construct an entire function $G_2(u)$ which has zeros at the poles of $f(u)$ and moreover whose zeros are of the same multiplicity as the corresponding poles of $f(u)$. Then the function $f(u)G_2(u) = G_1(u)$ is also entire, from which we get $f(u) = G_1(u)/G_2(u)$. The construction of $G_2(u)$ is made by means of an infinite product. However, this method is not applicable to a larger number of variables since the denominator of the fraction $\mathcal{M}_a(u_1, u_2)/\mathcal{N}_a(u_1, u_2)$ in the statement of the theorem of Poincaré is zero not at isolated points but on a manifold $N_a(u_1, u_2) = 0$ containing a.

The proof of the theorem of Poincaré was quite complicated and did not allow a natural generalization when the number of variables is more than 2. A simple and at the same time general (for any p) proof was given by the French mathematician P. Cousin in his doctoral thesis (1894). This proof serves as the basis for our exposition.

Everything is reduced to the following statement.

THEOREM OF COUSIN. *Associate with each point of the space $\mathbf{C}^p$ a function $g_a(u)$ which is analytic in some neighborhood $U(a)$ of this point. Also let the following coherence conditions hold for each point $b \in U(a^{(1)}) \cap U(a^{(2)})$, where $a^{(1)}$ and $a^{(2)}$ are different points of $\mathbf{C}^p$: $g_{a^{(1)}}(u)$ and $g_{a^{(2)}}(u)$ differ at this point only by a unit factor (i.e., $g_{a^{(2)}} | g_{a^{(1)}}$ and $g_{a^{(1)}} | g_{a^{(2)}}$ at the point b). Then there exists an entire function $G(u)$ such that*

$$G(u) = g_a(u) e_a(u), \tag{B.3.1}$$

where $e_a(u)$ is a unit at the point a (a is any point of $\mathbf{C}^p$).

The following theorem is a corollary.

THE POINCARÉ-COUSIN THEOREM. *If a function $f(u)$ can be represented in the form*

$$f(u) = \frac{\mathcal{M}_a(u)}{\mathcal{N}_a(u)},$$

in a neighborhood of each point $a \in \mathbf{C}^p$, where $\mathcal{M}_a(u)$ and $\mathcal{N}_a(u)$ are functions which are analytic at the point a, then there exist entire functions $G_1(u)$ and $G_2(u)$ such that

$$f(u) = \frac{G_1(u)}{G_2(u)}$$

at all points of $\mathbf{C}^p$. Moreover, we can require that the condition $(G_1(u), G_2(u)) = 1$ holds for each u.

Exactly this form of the theorem was used in Chapters II–IV of this book. We will show that it indeed follows from the Cousin theorem. We start by finding for each point $a \in \mathbf{C}^p$ the g.c.d of the functions $\mathcal{M}_n(a)$ and $\mathcal{N}_a(u)$ at that point and then we divide the fraction $\mathcal{M}_a(u)/\mathcal{N}_a(u)$ by this g.c.d. We obtain a local representation of $f(u)$ in the form of an irreducible fraction $f(u) = \mu_a(u)/\nu_a(u)$. Since it is irreducible at a point a, it is also irreducible at each point of some neighborhood $V(a)$ of the point a (Theorem B.2.1).

Now we apply Cousin's theorem to the case $g_a(u) = \nu_a(u)$. We check that the coherence condition holds here. Indeed, if $b \in V_{a^{(1)}} \cap V_{a^{(2)}}$, then in a sufficiently small neighborhood of this point we have

$$f(u) = \frac{\mu_{a^{(1)}}(u)}{\nu_{a^{(1)}}(u)} = \frac{\mu_{a^{(2)}}(u)}{\nu_{a^{(2)}}(u)},$$

from which we get

$$\mu_{a^{(1)}}(u)\nu_{a^{(2)}}(u) = \mu_{a^{(2)}}(u)\nu_{a^{(1)}}(u), \tag{B.3.2}$$

and moreover

$$(\mu_{a^{(1)}}(u), \nu_{a^{(1)}}(u))_b = 1 \quad \text{and} \quad (\mu_{a^{(2)}}(u), \nu_{a^{(2)}}(u))_b = 1. \tag{B.3.3}$$

Because of (B.3.2), by Euclid's theorem (section B.1) we have

$$\nu_{a^{(1)}}(u)|\nu_{a^{(2)}}(u) \quad \text{and} \quad \nu_{a^{(2)}}(u)|\nu_{a^{(1)}}(u),$$

at the point b and this is then the coherence condition. Hence there exists an entire function $G_2(u)$ which differs from $\nu_a(u)$ only by a unit factor $e_a(u) : G_2(u) = \nu_a(u)e_a(u)$ in a neighborhood of each point a. It follows from this that the function $f(u)G_2(u) = e_a(u)\mu_a(u)$ is an analytic function in a neighborhood of each point a and therefore it is an entire function $G_1(u)$.

It only remains to note that it follows from the equalities

$$G_1(u) = \mu_a(u)e_a(u), \qquad G_2(u) = \nu_a(u)e_a(u)$$

and from the condition $(\mu_a(u), \nu_a(u)) = 1$ that $(G_1, G_2) = 1$.

Thus everyting is reduced to the proof of Cousin's theorem. We precede it with two simple lemmas.

LEMMA 1. *Let C_n be the p-disk $|u_j| < n$, $j = 1, \dots, p$. In order to establish the truth of the Cousin theorem it suffices to prove that for each natural number n there exists a function $G_n(u)$ that is analytic on C_{n+1} and satisfies the conditions* (B.3.1) *at every point $a \in C_{n+1}$.*

PROOF. Assume that the conditions of the lemma hold. Then the function $G_{n+1}(u)/G_n(u)$ can be represented at each point $a \in C_{n+1}$ (by (B.3.1)) in the form of a ratio of two units and therefore is an analytic function, different from zero.

We define a single-valued analytic function $H_n(u)$ on C_{n+1} by

$$H_n(u) = \operatorname{Ln} \frac{G_{n+1}(u)}{G_n(u)}, \tag{B.3.4}$$

which is subject, for example, to the following condition at the point $u = 0$:

$$H_n(0) = \ln \frac{G_{n+1}(0)}{G_n(0)}.$$

This function has, in a neighborhood of the point $u = 0$, a power series expansion that converges uniformly on $\overline{C}_n$. Hence there exists a polynomial $P_n(u)$ (a partial sum of the series) such that

$$|H_n(u) - P_n(u)| < \frac{1}{n^2}, \quad u \in \overline{C}_n. \tag{B.3.5}$$

We show that the function

$$G(u) = G_1(u) \exp \sum_1^\infty [H_n(u) - P_n(u)] \tag{B.3.6}$$

is one of the entire functions mentioned in Cousin's theorem.

First, the series $\sum_N^\infty [H_n(u) - P_n(u)]$ converges uniformly in the p-disk C_N (by the conditions (B.3.5)) and therefore represents an analytic function on C_N (the theorem is proved by using Cauchy's integral formula for a p-disk in the same way as for one variable). Therefore the function (B.3.6) is at least analytic on C_1. But it can be continued analytically to C_N for any natural number N since (because of (B.3.4)) we have

$$\begin{aligned} G(u) &= \exp\left\{\left[L_n G_1(u) + \operatorname{Ln}\frac{G_2(u)}{G_1(u)} + \cdots + \operatorname{Ln}\frac{G_N(u)}{G_{N-1}(u)}\right]\right. \\ &\qquad \left. + \sum_N^\infty [H_n(u) - P_n(u)] - \sum_1^{N-1} P_n\right\} \\ &= G_N(u) \exp\left\{\sum_N^\infty [H_n(u) - P_n(u)] - \sum_1^{N-1} P_n(u)\right\}. \end{aligned} \tag{B.3.7}$$

Hence $G(u)$ is an entire function. Finally, the last formula shows that $G(u)$ differs from the function $G_N(u)$ only by a unit factor at each point $u \in C_N$ and therefore $G(u)$ also satisfies the conditions (B.3.1).

LEMMA 2. *Let l be a directed line segment with initial point a and end point b, let g be a region containing l, and let $h(z)$ be a function that is single-valued and analytic in this region. Then the integral*

$$I_h(z) = \frac{1}{2\pi i} \int_l \frac{h(\zeta)\, d\zeta}{\zeta - z} \tag{B.3.8}$$

of Cauchy type defines an analytic function on the z-plane, with the two branch points a and b. Moreover, $I_h(z)$ has the representation

$$I_h(z) = -\frac{1}{2\pi i} h(z) \operatorname{Ln}(a - z) + \varphi_1(z) \tag{B.3.9}$$

in a neighborhood of a and the representation

$$I_h(z) = \frac{1}{2\pi i} h(z) \operatorname{Ln}(b - z) + \varphi_2(z), \tag{B.3.10}$$

in a neighborhood of b, *where* $\varphi_1(z)$ *and* $\varphi_2(z)$ *are analytic at the corresponding points.*

PROOF. It is obvious that (B.3.8) is a single-valued analytic branch of some function on the plane with a cut along l. In a neighborhood of l, i.e., for $z \in g$, $I_h(z)$ can be represented in the form

$$I_h(z) = \frac{1}{2\pi i} h(z) \operatorname{Ln} \frac{b-z}{a-z} + \frac{1}{2\pi i} \int_l \frac{h(\zeta) - h(z)}{\zeta - z} \, d\zeta, \qquad \text{(B.3.11)}$$

from which the assertion of the lemma follows, since the function defined by the integral on the right-hand side is regular at each point of l. It also follows from this that the single-valued branches which are obtained by analytic continuation of the function across the cut differ from each other at the points of g by integral multiples of $h(z)$. In fact, when z moves from the left edge to the right edge (i.e., when z goes around the point a in the positive direction) the function increases by $h(z)$.

Now we can establish Cousin's theorem. We replace the p-disks C_n of Lemma 1 by p-disks of the form $D_n : |x_j| < n$, $|y_j| < n$, $j = 1, \dots, p$, where x_j and y_j are the real and imaginary parts of u_j. Since $C_n \subset D_n$ for any n, the conclusion of the lemma still holds.

PROOF OF COUSIN'S THEOREM. We associate with each point, not only the neighborhood $U(a)$: $\{|u_j - a_j| < r_a, j = 1, \dots, p\}$ which appeared in the hypotheses of the theorem, but also a concentric neighborhood $V_a(u)$: $\{|u_j - a_j| < \frac{1}{2} r_a\}$ of radius $r_a/2$. With n fixed, we cover $\overline{D}_{n+1}$ by a finite number of neighborhoods $V_1, \dots, V_n$ with centers at the points $a^{(1)}, \dots, a^{(N)}$. We denote the corresponding analytic functions $g_a(u)$ by $g_1(u), \dots, g_N(u)$. If $r = r_n = \min\{r_{a^{(1)}}, \dots, r_{a^{(N)}}\}$, we dissect the large squares $T_j : \{|x_j| < n + 1, |y_j| < n + 1\}$, $j = 1, \dots, p$, into equal squares, with sides parallel to the coordinate axes, so small that the diagonal of each one is less than r. Then the region D_{n+1} is broken into a finite number of elementary parallelotopes each of which is the topological product of p small squares taken from the p partitions T_j $(j = 1, \dots, p)$. If at least one point of a parallelotope τ belongs to a neighborhood V_ν, then the entire parallelotope will be inside U_ν: $|u_j - a_j^{(\nu)}| < r_{a^{(\nu)}}$, $j = 1, \dots, p$, and therefore the corresponding function $g_\nu(u)$ will be defined and analytic inside τ and on its boundary. Of course, τ can be contained in some other neighborhood U_μ related to the function $g_\mu(u)$. However, because of the coherence condition we have $g_\nu(u) | g_\mu(u)$ and $g_\mu(u) | g_\nu(u)$ at all points of $\bar{\tau}$ (i.e., inside τ and on its boundary).

We also notice that if σ and τ are different parallelotopes which have at least one common boundary point b, and if the point b belongs to the neighborhood V_ν, then it follows that σ and τ are contained entirely in the same U_ν and hence the same function $g_\nu(u)$ is defined at σ and τ.

We now need to construct a function $G_n(u)$ that is analytic in the p-disk D_{n+1} and satisfies the conditions (B.3.1) at all points of D_{n+1}. This construction starts from a finite number of functions $g_\nu(u)$, $\nu = 1, \dots, N$; a significant role in this construction is played by Lemma 2. The easiest way to make this construction is by induction on the number of variables.

For each elementary parallelotope τ we select one of the functions $g_\nu(u)$ which are defined on it, and denote this function by $g_\tau(u)$. Then we will have a locally analytic function $\Phi(u)$ on the region D_{n+1}. This function is single-valued and analytic inside each τ, where $\Phi(u) = g_\tau(u)$. As for boundary points common to two or more parallelotopes, the function is multiple-valued at these points. However, if, for example, a point b is common to τ and σ, then there is a neighborhood of this point in which the coherence condition $g_\tau(u)|g_\sigma(u)$ and $g_\sigma(u)|g_\tau(u)$ holds. For brevity, we denote such a relation by $g_\tau(u) \sim g_\sigma(u)$ and say that $g_\tau(u)$ and $g_\sigma(u)$ are *equivalent* (at the point b).

We separate the variables $u_1, \dots, u_p$ into two groups A and B: $u_{i_1}, \dots, u_{i_k}$ and $u_{j_1}, \dots, u_{j_l}$ $(k + l = p)$ in such a way that each variable belongs to one and only one group. The inductive assumption will be that the *locally analytic* function $\Phi(u)$ is single-valued and analytic at each point $u \in D_{n+1}$ with respect to all variables of the group A when arbitrary fixed values are assigned to the variables of the group B (the group A may initially be empty). We will show that we can move from $\Phi(u)$ to another function Φ_1 which is equivalent to Φ at all points D_{n+1}, in such a way that we add another variable, for instance $u_{j_1} = v$, of group B to group A.

With this goal in mind, we arbitrarily fix the values of all variables of the group A, as well as of group B with the exception of v. Then we obtain a function of one complex variable v which we still denote by $\Phi(v)$ and which is locally analytic on the square $T_{j_1} = T$. This square was decomposed into small squares t. In each of these squares, $\Phi(v)$ coincides with the corresponding single-valued analytic function $g_t(v)$. Let t_α and t_β be two squares which have a common side $l_{\alpha\beta}$. We will orient $l_{\alpha\beta}$ so that the square t_α lies to the left of it and the square t_β lies to the right of it. Then at all of its points the functions $g_\alpha(v) = g_{t_\alpha}(v)$ and $g_\beta(v) = g_{t_\beta}(v)$ are analytic and equivalent. Hence $g_\beta(v)/g_\alpha(v)$ is also an analytic zero-free function of v. Therefore in some region $g_{\alpha\beta}$ which contains $l_{\alpha\beta}$ the function

$$h_{\alpha\beta} = \operatorname{Ln} \frac{g_\beta(v)}{g_\alpha(v)} \tag{B.3.12}$$

is determined up to an integral multiple of $2\pi i$, and is single-valued and analytic.

By Lemma 2 the function

$$I_{\alpha\beta}(v) = \frac{1}{2\pi i} \int_{l_{\alpha\beta}} \frac{h_{\alpha\beta}(\zeta)\, d\zeta}{\zeta - v} \tag{B.3.13}$$

is then single-valued and analytic on T with a cut along $l_{\alpha\beta}$. In general, it has singularities α and β. We form the sum

$$\Lambda(v) = \sum I_{\alpha\beta}(v), \tag{B.3.14}$$

by extending it to all pairs of squares t_α and t_β which have a common side. The function $\Lambda(v)$ is analytic inside each square t_α. When we extend it analytically from t_α to t_β, which has a common side $l_{\alpha\beta}$ with t_α, the function gains a summand $h_{\alpha\beta}$, by the lemma. We now select the multiples of $2\pi i$ which are included in the definition (B.3.12) in such a way that at each of the common vertices b of the four squares t_α, t_β, t_γ, and t_δ

$$\begin{aligned} &h_{\alpha\beta}(b) + h_{\beta\gamma}(b) + h_{\gamma\delta}(b) + h_{\delta\alpha}(b) \\ &\quad = \ln \frac{g_\beta(b)g_\gamma(b)g_\delta(b)g_\alpha(b)}{g_\alpha(b)g_\beta(b)g_\gamma(b)g_\delta(b)} + 2\pi i(n_{\alpha\beta} + n_{\beta\gamma} + n_{\gamma\delta} + n_{\delta\alpha}) \\ &\quad = 0. \end{aligned} \tag{B.3.15}$$

It is obvious that all this is possible if, for example, we traverse the squares which belong to T sequentially along the columns from the bottom up and move from a completed column to the next, from left to right. The choice of the multiples $2\pi i$ for the squares of the first column and for the bottom square of the second column can be arbitrary. However, the choice for further steps is forced uniquely by (B.3.15).

Under (B.1.15) the following representation (Lemma 2) exists in a neighborhood of each common vertex b of the four squares :

$$I_{\alpha\beta}(v) + I_{\beta\gamma}(v) + I_{\gamma\delta}(v) + I_{\delta\alpha}(v) = \varphi(v),$$

where $\varphi(v)$ is analytic at b.

This means that the function $\Lambda(v)$ is analytic at such a vertex. We will verify that the function

$$\Phi_1(v) = \Phi(v)\exp[\Lambda(v)] \tag{B.3.16}$$

has the inductive properties needed to complete the proof. Indeed, at the points of each closed square $\bar{t}_\alpha \subset T$ we have $\Phi_1(v) \sim \Phi(v) = g_\alpha(v)$. Furthermore, it is single-valued and analytic at each boundary point of the square t_α since the exponent $\Lambda(v)$, i.e., the function (B.3.14), gains $h_{\alpha\beta}(v) = \ln(g_\beta(v)/g_\alpha(v))$ from analytic continuation across $l_{\alpha\beta}$ to the adjacent square t_β. Consequently $\Phi_1(v)$ gains the factor $g_\beta(v)/g_\alpha(v)$ and thus, in the square t_β,

$$\Phi_1(v) = g_\beta(v)\exp[\Lambda(v)] = \Phi(v)\exp[\Lambda(v)], \tag{B.3.16$'$}$$

which coincides with the definition of $\Phi_1(v)$ on this square.

Since the inductive assumption is trivially true when the group of variables A is empty, the proof of Cousin's theorem is complete. We proved by induction that there exists a function $G_n(u)$, analytic in D_{n+1} and satisfying at each point of this region the condition $G_n(u) \sim g_\alpha(u)$. From this,

Cousin's theorem follows by Lemma 1. Hence the Poincaré-Cousin theorem is established.

§C. A summary of the most important formulas

Jacobi function.

$$\vartheta_1(x) = \sum_0^\infty (-1)^n 2q^{(2n+1)^2/4} \sin(2n+1)\pi x,$$

$$\vartheta_2(x) = \sum_0^\infty 2q^{(2n+1)^2/4} \cos(2n+1)\pi x,$$

$$\vartheta_3(x) = 1 + \sum_1^\infty 2q^{n^2} \cos 2\pi x,$$

$$\vartheta_4(x) = 1 + \sum_1^\infty (-1)^n 2q^{n^2} \cos 2n\pi x.$$

$$\operatorname{sn} u = \frac{1}{\sqrt{k}} \frac{\vartheta_1(u/(2K))}{\vartheta_4(u/(2K))}, \qquad \operatorname{cn} u = \sqrt{\frac{k'}{k}} \frac{\vartheta_2(u/(2K))}{\vartheta_4(u/(2K))},$$

$$\operatorname{dn} u = \sqrt{k'} \frac{\vartheta_3(u/(2K))}{\vartheta_4(u/(2K))}.$$

$$k^2 + k'^2 = 1, \qquad K = \int_0^{\pi/2} \frac{d\varphi}{\sqrt{1 - k^2 \sin^2 \varphi}}, \qquad K' = \int_0^{\pi/2} \frac{d\varphi}{\sqrt{1 - k'^2 \sin^2 \varphi}}.$$

The Weierstrass preparation theorem. If $f(u) = \Sigma_{n=\nu}^\infty p_n(u)$, where $p_n(u)$ are homogeneous polynomials of degree n and $p_\nu(u) = \pi_0 u^\nu + \cdots + \pi_\nu$, $\pi_0 \neq 0$, $\nu \geq 1$, then

$$f(u) = (\Pi_0 u_1^\nu + \cdots + \Pi_\nu) f_1(u);$$

here $\Pi_0 = \text{const} \neq 0$, $\Pi_j = \Pi_j(u')$, the lower order of Π_j is not less than j $(j = 1, \ldots, \nu)$, and $f_1(0) \neq 0$.

Reduced representations. If $f(u) = \varphi(u)/\psi(u) = \varphi^*(u)/\psi^*(u)$ are two reduced representations of a meromorphic function, then there is an entire function $g(u)$ such that $\varphi^*(u) = \exp[g(u)]\varphi(u)$, $\psi^*(u) = \exp[g(u)]\psi(u)$.

A condition for the degeneracy of a function. A function $f(u)$ is degenerate if and only if a p-dimensional vector $\tilde{\gamma} \neq 0$, $\tilde{\gamma} = (\gamma_1, \ldots, \gamma_p)$, exists such that

$$\gamma_1 \frac{\partial f}{\partial u_1} + \cdots + \gamma_p \frac{\partial f}{\partial u_p} = 0$$

at each regular point of the function.

Period matrices. If $\Omega = (\omega^{(1)}, \ldots, \omega^{(r)})$ and $\Pi = (\pi^{(1)}, \ldots, \pi^{(\rho)})$ are two matrices of fundamental periods of the function $f(u)$, then $\rho = r$ and there exists an (r, r) matrix M with integer entries such that $|M| = \pm 1$ (is *unimodular*) and $\Pi = \Omega M^{*}$.

If the variables $u_1, \dots, u_p$ are subjected to a linear transformation $u^* = Au$, $|A| \neq 0$, then the period matrix Ω transforms into $\Omega^* = A\Omega$.

Fourier series expansion. If each column of the matrix $C = \mathrm{Diag}(c_1, \dots, c_p)$, $c_j \neq 0$, $j = 1, \dots, p$, is a period of an entire function $f(u)$, then

$$f(u) = \sum_{(\gamma)} d_\gamma \exp(2\pi i \tilde{\gamma} C^{-1} u),$$

where the summation is over all p-vectors γ with integral coordinates.

Construction of an entire function from its difference. If $f(u)$ is an entire function of one variable $\tau \neq 0$ and

$$f(u + \tau) - f(u) = h(u),$$

where $h(u) = \sum_0^\infty h_n u^n$ is an entire function, then

$$f(u) = \sum_0^\infty h_n \tau^n \psi_n(u/\tau) + P(u),$$

where

$$\psi_n(u) = \frac{n!}{2\pi i} \int_{|z|=(2n+1)\pi} \frac{e^{uz} - 1}{e^z - 1} \frac{dz}{z^{n+1}}, \qquad n = 0, 1, 2, \dots$$

and $P_n(u)$ is any entire function with period τ.

Riemann matrices. Elementary conditions. If $\Omega = (\omega^{(1)}, \dots, \omega^{(2p)})$ is a Riemann matrix and $\Omega' = \mathrm{Re}\,\Omega$ and $\Omega'' = \mathrm{Im}\,\Omega$, then

$$D = \begin{vmatrix} \Omega' \\ \Omega'' \end{vmatrix} \neq 0 \quad \text{and therefore} \quad |F| = \begin{vmatrix} \Omega \\ \bar{\Omega} \end{vmatrix} = (-2i)^p D \neq 0.$$

Each equivalence class of Riemann matrices contains a matrix of the form

$$\Omega = (2\pi i E_p, T).$$

Here $|\mathrm{Re}\,T| \neq 0$. Furthermore, if $T = (t^{(1)}, \dots, t^{(p)})$, there exists a number $\alpha = \alpha(T) > 0$ such that we have, for each p-vector γ with integral coefficients,

$$\max_{1 \le k \le p} |\exp[\tilde{\gamma} t^{(k)}] - 1| \ge \alpha.$$

The first system of difference equations. If $f(u) = \varphi(u)/\psi(u)$ is an Abelian function with matrix $\Omega = (2\pi i E_p, T)$, then a reduced representation

$$f(u) = \frac{\varphi(u) \exp G(u)}{\psi(u) \exp G(u)} = \frac{\varphi^*(u)}{\psi^*(u)}$$

exists such that

$$\varphi(u + 2\pi i e^{(h)}) = \exp[l_h(u)]\varphi(u),$$

$$\psi(u + 2\pi i e^{(h)}) = \exp[l_h(u)]\psi(u),$$

$$h = 1, \dots, p,$$

where $l_h(u)$ are linear functions.

An entire function $G(u)$ satisfies the (first) system of difference equations

$$G(u + 2\pi i e^{(h)}) - G(u) = H_h(u), \qquad h = 1, \dots, p,$$

where $H_h(u)$ are entire functions for which

$$H_h(u + 2\pi i e^{(k)}) - H_h(u) = H_k(u + 2\pi i e^{(h)}) - H_k(u),$$
$$h, k = 1, \dots, p.$$

The solution $G(u)$ of the first system is determined up to an additive entire function $P(u)$ with periods $2\pi i e^{(k)}$, $k = 1, \dots, p$.

Jacobian functions. An entire function $\varphi(u)$ is called *Jacobian* with respect to a Riemann matrix Ω if for any period ω generated by Ω we have $\varphi(u + \omega) = \exp[\lambda(u) + c]\varphi(u)$, where $\lambda(u) = \lambda_\omega(u)$ is a linear function and $c = c_\omega$ is a constant.

Let $\Omega = (2\pi i E_p, T)$. In order to represent an Abelian function $f(u) = \varphi^*(u)/\psi^*(u)$ as a quotient of two Jacobian functions we have to choose a periodic function $P(u)$ in the expression $G(u) = G_0(u) + P(u)$, where $G_0(u)$ is some solution of the first system, and choose it in such a way that the *second* system

$$P(u + t^{(h)}) - P(u) = \widetilde{H}_h(u), \qquad h = 1, \dots, p,$$

is satisfied. Here $\widetilde{H}_h(u)$ are entire functions that satisfy the conditions

$$\widetilde{H}_h(u + t^{(k)}) - \widetilde{H}_h(u) = \widetilde{H}_k(u + t^{(h)}) - \widetilde{H}_k(u), \qquad h, k = 1, \dots, p.$$

The function $P(u)$ is defined in the form of a Fourier series, up to an additive constant.

Necessary and sufficient conditions for Ω. A $(p, 2p)$ matrix Ω is a Riemann matrix if and only if there exists a *principal* matrix for it, where the principal matrix P is a $(2p, 2p)$ matrix with integral entries such that

$$\widetilde{P} = -P, \qquad \Omega P \widetilde{\Omega} = 0, \qquad i\Omega P \widetilde{\overline{\Omega}} > 0 \ \ (\text{or} \ < 0).$$

Let $\varphi(u)$ be a Jacobian function and let

$$\varphi(u + \omega^{(k)}) = \varphi(u + \widetilde{\varepsilon}^{(k)}\Omega) = \exp[2\pi i \widetilde{\varepsilon}^{(k)}(\widetilde{\Lambda} u + \gamma)], \qquad k = 1, \dots, 2p.$$

where $\varepsilon^{(k)}$ is the kth column of a matrix E_{2p} and γ is a $2p$-vector (*a parametric vector* of the function $\varphi(u)$), and let

$$\Lambda = (\lambda_{jk}), \qquad j, k = 1, \dots, p,$$

be the *second period matrix.* Then

$$N = \tilde{\Omega}\Lambda - \tilde{\Lambda}\Omega$$

is the *characteristic matrix.* Its entries are integers which do not depend on the choice of the function $\varphi(u)$. Then nN^{-1} can be taken as the principal matrix for P, where n is an integer.

Theta function. The matrices Ω, N, and Λ can be expressed in the form

$$\Omega = (\pi i B^{-1}, A), \qquad N = n\begin{pmatrix} 0 & B^{-1} \\ B^{-1} & 0 \end{pmatrix}, \qquad \Lambda = \left(0, -\frac{n}{\pi i}E_p\right),$$

where A is a symmetric (p, p) matrix such that $\operatorname{Re} A = A' < 0$ and $B = \operatorname{Diag}(\beta_1, \dots, \beta_p)$, $nB^{-1} = \operatorname{Diag}(\delta_p, \dots, \delta_1)$. The natural numbers β_j, δ_u, l, and n satisfy the following conditions: $1 = \beta_1|\beta_2|\cdots|\beta_p$, $\delta_k = \beta_p/\beta_{p-k+1}$, $1 = \delta_1|\delta_2|\cdots|\delta_p = \beta_p$, $n = l\beta_p = l\delta_p$, $l = 1, 2, \dots$.

We also choose a parametric vector γ in such a way that

$$\tilde{\gamma} = (0, -\frac{n}{2\pi i}\operatorname{Sp} A),$$

where the p-vector $\operatorname{Sp} A$ is formed by the diagonal entries of A. Then any Jacobian function of the type $\{\Omega, \Lambda, \gamma\}$ can be represented as a linear combination of *theta functions of order* n:

$$\Theta_n[g](u) = \sum_m \exp\left\{\left(\tilde{m} + \frac{1}{n}\tilde{g}B\right) n \left[A\left(m + \frac{1}{n}Bg\right) + 2u\right]\right.$$

Here g is an arbitrary vector with integral coordinates which belongs to a parallelotope Π constructed from the vectors $(n/\beta_h)e^{(h)}$, $h = 1, \dots, p$. The summation is taken over all p-vectors m with integral coordinates.

The number of different functions $\Theta_n[g](u)$ is equal to $\sqrt{|N|}$, and they are linearly independent.

Algebraic dependencies. Any $p + 2$ Jacobian functions of p variables of the same type are related by an algebraic relation (homogeneous relation).

Any $p + 1$ Abelian functions of the same field are related by an algebraic relation.

In any field of Abelian functions there exist p functions which are algebraically, and even analytically, independent.